Y0-CAR-399

AN INTRODUCTION TO PREDICTIVE MAINTENANCE

AN INTRODUCTION TO PREDICTIVE MAINTENANCE

R. Keith Mobley
Knoxville, Tennessee

Plant Engineering Series

Van Nostrand Reinhold
New York

Copyright © 1990 by Van Nostrand Reinhold
Library of Congress Catalog Card Number 89-9150
ISBN 0-442-31828-6

Printed in the United States of America

Van Nostrand Reinhold
115 Fifth Avenue
New York, New York 10003

Van Nostrand Reinhold International Company Limited
11 New Fetter Lane
London EC4P 4EE, England

Van Nostrand Reinhold
480 La Trobe Street
Melbourne, Victoria 3000, Australia

Nelson Canada
1120 Birchmount Road
Scarborough, Ontario M1K 5G4, Canada

16 15 14 13 12 11 10 9 8 7 6 5 4 3 2 1

Library of Congress Cataloging-in-Publication Data

Mobley, R. Keith, 1943–
    An introduction to predictive maintenance / written by R. Keith
Mobley.
        p. cm. — (VNR plant engineering series)
    Includes index.
    ISBN 0-442-31828-6
    1. Plant maintenance—Management. I. Title. II. Series.
TS192.M624 1989
658.2′02—dc20
                                                    89-9150
                                                        CIP

Van Nostrand Reinhold's Plant Engineering Series

Preface

Predictive maintenance management has been widely discussed over the past few years. A variety of techniques ranging from vibration monitoring to infrared imaging has been defined as the best and often the only technology suitable for implementing a program. New terminology is now surfacing that is represented as being the new maintenance management tool that should be used to run your maintenance operation. These new terms—RCM, reliability-centered maintenance; TPM, total productive maintenance; and JIT, just-in-time maintenance—are presented as replacements for predictive maintenance and the ultimate solution to your high maintenance costs.

How do you decide which of these maintenance management techniques should be used in your plant? How do you select the best and most cost-effective techniques? How do you implement a program that will be successful?

This book is designed to provide the practical knowledge that you will need to select and implement a comprehensive, cost-effective maintenance management program in your plant. More importantly, it is designed to provide the methods and implementation steps that will ensure that your program will be successful. The information provided has successfully helped hundreds of manufacturing and process plants worldwide.

Since most manufacturing and process plants rely on mechanical equipment for most of their process, vibration-based predictive maintenance is the dominant technique used for most maintenance management programs. *However, the ability to monitor all critical machines, equipment, and systems in a typical plant cannot be limited to a single technique.*

Predictive, or condition-based, monitoring techniques include:

vibration analysis, ultrasonics, thermography, tribology, process monitoring, visual inspection, and other nondestructive analysis techniques. This combination of monitoring and analysis techniques provides the means of directly monitoring all of the critical equipment and systems in your plant.

Contents

AN INTRODUCTION TO PREDICTIVE MAINTENANCE

Chapter 1

Introduction

Maintenance costs are a major part of the total operating costs of all manufacturing and production plants. Depending on the specific industry, maintenance costs can represent between 15% to 40% of the costs of goods produced. For example, in food-related industries, the average maintenance costs represent about 15% of the cost of goods produced; while in iron and steel, pulp and paper, and other heavy industries, maintenance represents up to 40% of the total production costs.

Recent surveys of maintenance management effectiveness indicate that one third of all maintenance costs is wasted as the result of unnecessary or improperly carried out maintenance. When you consider that United States industry spends more than 200 billion dollars each year on maintenance of plant equipment and facilities, the impact on productivity and profit that is represented by the maintenance operation becomes clear.

The result of ineffective maintenance management represents a loss of more than 60 billion dollars each year. Perhaps more important is the fact that our ineffective management of maintenance dramatically impacts our ability to manufacture quality products that are competitive in the world market. The loss of production time and product quality that results from inadequate maintenance management has had a dramatic impact on our ability to compete with Japan and other countries that have implemented more advanced manufacturing and maintenance management philosophies.

The dominant reason for this ineffective management is the lack of factual data that quantifies the actual need for repair or mainte-

1

nance of plant machinery, equipment, and systems. Maintenance scheduling has been and, in many instances, is predicated on statistical trend data or on the actual failure of plant equipment.

Until recently, middle and corporate-level management have ignored the impact of the maintenance operation on product quality, production costs, and, more importantly, on bottom-line profit. The general opinion has been "Maintenance is a necessary evil," or "Nothing can be done to improve maintenance costs." Perhaps these were true statements 10 or 20 years ago.

However, the development of microprocessor and other computer-based instrumentation used to monitor the operating condition of plant equipment, machinery, and systems has provided the means to manage the maintenance operation. They have enabled staff to reduce or eliminate unnecessary repairs, prevent catastrophic machine failures, and reduce the negative impact of the maintenance operation on the profitability of manufacturing and production plants.

To understand predictive maintenance management programs, traditional management techniques should first be considered. Industrial and process plants typically utilize two types of maintenance management: run-to-failure or preventive maintenance.

RUN-TO-FAILURE MANAGEMENT

The logic of run-to-failure management is simple and straightforward: when a machine breaks down, fix it. This "If it ain't broke, don't fix it" method of maintaining plant machinery has been a major part of plant maintenance operations since the first manufacturing plant was built and, on the surface, sounds reasonable. A plant using run-to-failure management does not spend any money on maintenance until a machine or system fails to operate.

Run-to-failure is a reactive management technique that waits for machine or equipment failure before any maintenance action is taken. It is in truth a "no maintenance" approach to management. It is also the most expensive method of maintenance management.

Few plants use a true run-to-failure management philosophy. In almost all instances, plants perform basic preventive tasks, such as lubrication and machine adjustments, even in a run-to-failure environment. However, in this type of management, machines and other plant equipment are not rebuilt nor are any major repairs made until the equipment fails to operate.

The major expenses associated with this type of maintenance management are: high spare parts inventory cost, high overtime labor costs, high machine downtime, and low production availability.

Since there is no attempt to anticipate maintenance requirements, a plant that uses true run-to-failure management must be able to react to all possible failures within the plant. This reactive method of management forces the maintenance department to maintain extensive spare parts inventories that include spare machines or at least all major components for all critical equipment in the plant. The alternative is to rely on equipment vendors who can provide immediate delivery of all required spare parts.

Even if the latter is possible, premiums for expedited delivery substantially increase the costs of repair parts and downtime required to correct machine failures. To minimize the impact on production created by unexpected machine failures, maintenance personnel must also be able to react immediately to all machine failures. The net result of this reactive type of maintenance management is higher maintenance cost and lower availability of process machinery. Analysis of maintenance costs indicates that a repair performed in the reactive run-to-failure mode will average about three times higher than the same repair made within a scheduled or preventive mode. Scheduling the repair provides the ability to minimize the repair time and associated labor costs. It also provides the means of reducing the negative impact of expedited shipments and lost production.

PREVENTIVE MAINTENANCE MANAGEMENT

There are many definitions of preventive maintenance. However, all preventive maintenance management programs are time-driven. In other words, maintenance tasks are based on elapsed time or hours of operation. Figure 1-1 illustrates an example of the statistical life of a machine-train. The mean-time-to-failure (MTTF) or bathtub curve indicates that a new machine has a high probability of failure, due to installation problems, during the first few weeks of operation. Following this initial period, the probability of failure is

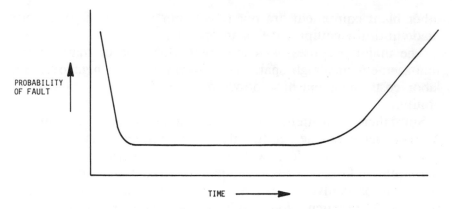

Figure 1-1. Bathtub curve illustrates the life cycle of a specific classification of machinery.

relatively low for an extended period of time. Following this normal machine life period, the probability of failure increases sharply with elapsed time. In preventive maintenance management, machine repairs or rebuilds are scheduled based on the MTTF statistic.

The actual implementation of preventive maintenance varies greatly. Some programs are extremely limited and consist of lubrication and minor adjustments. More comprehensive preventive maintenance programs schedule repairs, lubrication, adjustments, and machine rebuilds for all critical machinery in the plant. The common denominator for all of these preventive maintenance programs is the scheduling guideline, time.

All preventive maintenance management programs assume that machines will degrade within a time frame typical of its particular classification. For example, a single-stage, horizontal, split-case centrifugal pump will usually run 18 months before it must be rebuilt. Using preventive management techniques, the pump would be removed from service and rebuilt after 17 months of operation.

The problem with this approach is that the mode of operation and system- or plant-specific variables directly affect the normal operating life of machinery. The mean-time-between-failures (MTBF) will not be the same for a pump that is handling water and one handling abrasive slurries. The normal result of using MTBF statistics to schedule maintenance is either an unnecessary repair or a catastrophic failure. In the example, the pump may not need to be rebuilt after 17 months. Therefore, the labor and material used to make the repair were wasted. The second scenario of preventive maintenance is even more costly. If the pump fails before 17 months, we are forced to repair using run-to-failure techniques. Analysis of maintenance costs have shown that a repair made in a reactive (i.e., after failure) mode will normally be three times more expensive than the same repair made on a scheduled basis, for the reasons cited in the previous section.

The old adage that machines will break down at the worst possible time is a very real part of plant maintenance. Usually the breakdown will occur when production demands are the greatest. Maintenance staff must then react to the unexpected failure. In this mode of reactive maintenance, the machine is disassembled and inspected to determine the specific repairs that are required to return it to service. If repair parts are not in inventory, they must be ordered, at premium costs, and expedited shipment requested.

Even when the repair parts are already in plant inventory, the repair labor time and cost are much greater in this kind of reactive maintenance. The maintenance crew must disassemble the entire machine to locate the source of the problem or problems that forced the failure. Assuming that they correctly identify the problem, the time required to disassemble, repair, and reassemble the machine would be at least three times longer than would have been required by a planned repair.

In predictive maintenance programs, the specific failure mode (i.e., problem) can be identified *before* failure. Therefore, the correct repair parts, tools, and labor skills can be available to correct the machine problem before catastrophic failure occurs.

Perhaps the most important difference between reactive and predictive maintenance is the ability to schedule the repair when it will have the least impact on production. The production time lost as a result of reactive maintenance is substantial and can rarely be regained. Most plants, during peak production periods, operate 24 hours per day. Therefore, lost production time cannot be recovered.

PREDICTIVE MAINTENANCE

Like preventive maintenance, predictive maintenance has many definitions. To some, predictive maintenance is monitoring the vibration of rotating machinery in an attempt to detect incipient problems and to prevent catastrophic failure. To others, it is monitoring the infrared images of electrical switch gears, motors, and other electrical equipment to detect developing problems.

The common premise of predictive maintenance is that regular monitoring of the actual mechanical condition, operating efficiency, and other indicators of the operating condition of machine-trains and process systems will provide the data required to ensure the maximum interval between repairs. It would also minimize the number and costs of unscheduled outages created by machine-train failures.

Predictive maintenance is much more. It is a means of improving productivity, product quality, profitability, and overall effectiveness of our manufacturing and production plants. Predictive maintenance is not merely vibration monitoring or thermal imaging or lubricating oil analysis or any of the other nondestructive testing techniques that are being marketed as predictive maintenance tools. Predictive maintenance is a philosophy or attitude that uses the actual operating condition of plant equipment and systems to optimize total plant operation. A comprehensive predictive maintenance management program utilizes a combination of the most cost-effective tools to obtain the actual operating condition of critical plant systems, and, based on these actual data, all maintenance activities are scheduled on an "as-needed" basis.

Predictive maintenance is a condition-driven preventive maintenance program. Instead of relying on industrial or in-plant average-life statistics (e.g., mean-time-to-failure) to schedule maintenance activities, predictive maintenance uses direct monitoring of the mechanical condition, system efficiency, and other indicators to determine the actual mean-time-to-failure or loss of efficiency for each machine-train and system in the plant. At best, traditional time-driven methods provide a guideline to "normal" machine-train life spans.

In preventive or run-to-failure programs, the final decision on repair or rebuild schedules is based on the intuition and personal experience of the maintenance manager. The addition of a comprehensive predictive maintenance program can provide data on the actual mechanical condition of each machine-train and the operating efficiency of each process system. These data will enable the maintenance manager to schedule maintenance activities much more cost-effectively.

A predictive maintenance program can minimize the number of breakdowns of all mechanical equipment in the plant and ensure that repaired equipment is in acceptable mechanical condition. It can identify machine-train problems before they become serious since most mechanical problems can be minimized if they are detected and repaired early. Normal mechanical failure modes degrade at a speed directly proportional to their severity; therefore, when a problem is detected early, major repairs can usually be prevented.

There are five nondestructive techniques that are normally used for predictive maintenance management: vibration monitoring, process parameter monitoring, thermography, tribology, and visual inspection. Each technique has a unique data set that will assist the maintenance manager in determining the actual need for maintenance.

Predictive maintenance utilizing vibration signature analysis is predicated on two basic facts: (1) all common failure modes have distinct vibration frequency components that can be isolated and identified, and (2) the amplitude of each distinct vibration component will remain constant unless there is a change in the operating dynamics of the machine-train. These facts, their impact on machinery, and the methods used to identify and quantify the root cause of failure modes will be developed in more detail in later chapters.

Predictive maintenance utilizing process efficiency, heat loss, or other nondestructive techniques can quantify the operating efficiency of nonmechanical plant equipment or systems. These techniques used in conjunction with vibration analysis can provide the maintenance manager or plant engineer with factual information that will enable them to achieve optimum reliability and availability from their plant.

How do you determine which technique or techniques are required in your plant? How do you determine the best method to implement each of the technologies? If you listen to the salespeople for the vendors that supply predictive maintenance systems, theirs is the only solution to your problem. How do you separate the good from the bad?

Most comprehensive predictive maintenance programs will use vibration analysis as the primary tool. Since the majority of normal plant equipment is mechanical, vibration monitoring will provide the best tool for routine monitoring and identification of incipient problems. However, *vibration analysis will not provide the data required on electrical equipment, areas of heat loss, condition of lubricating oil, or other parameters that should be included in your program.*Therefore, a *total plant* predictive maintenance program must include several techniques, each designed to provide specific information on plant equipment, to achieve the benefits that this type of maintenance management can provide.

The specific techniques will depend on the type of plant equipment, their impact on production and on other key parameters of plant operation, and the goals and objectives that you want the predictive maintenance program to achieve.

Chapter 2

Benefits of Predictive Maintenance

Predictive maintenance is not a substitute for the more traditional maintenance management methods. It is, however, a valuable addition to a comprehensive, *total plant* maintenance management program. Whereas traditional maintenance management programs rely on routine servicing of all machinery and fast response to unexpected failures, a predictive maintenance program schedules specific maintenance tasks as they are actually required by plant equipment. It cannot totally eliminate all aspects of the traditional run-to-failure and preventive programs, yet predictive maintenance can reduce the number of unexpected failures as well as provide a more reliable scheduling tool for routine preventive maintenance tasks.

The premise of predictive maintenance is that regular monitoring of the actual mechanical condition of machine-trains and the operating efficiency of process systems will ensure the maximum interval between repairs. It will also minimize the number and cost of unscheduled outages created by machine-train failures and improve the overall availability of operating plants. Including predictive maintenance in a total plant management program will provide the ability to optimize the availability of process machinery and greatly reduce the cost of maintenance. In reality, predictive maintenance is a condition-driven preventive maintenance program.

A 1988 survey of 500 plants that have successfully implemented predictive maintenance methods indicates substantial improvements in reliability, availability, and operating costs. Conducted by the Plant Performance Group (a division of Technology for Energy Corporation), this survey was designed to quantify the impact of

9

including predictive maintenance techniques as a key part of the maintenance management philosophy. The sample group included a variety of industries in the United States, Canada, Great Britain, France, and Australia. The industries included electric power generation, pulp and paper, food processing, textiles, iron and steel, aluminum, and other manufacturing or process industries. Each of the participants had an established predictive maintenance program with a minimum of three years of program implementation.

The successful programs included in the survey provide an overview of the types of improvements that can be expected from a comprehensive predictive maintenance management program (see Figure 2-1). According to the survey results, major improvements can be achieved in maintenance costs, unscheduled machine failures, repair downtime, spare parts inventory, and both direct and indirect overtime premiums. In addition, the survey indicated a

WHAT ARE THE BENEFITS

OF A COMPREHENSIVE PREDICTIVE

MAINTENANCE PROGRAM ?

- MAINTENANCE COSTS REDUCED BY 50–80%

- MACHINE BREAKDOWN REDUCED BY 50–60%

- SPARE PARTS INVENTORY REDUCED BY 20–30%

- MACHINE DOWNTIME REDUCED BY 50–80%

- OVERTIME PREMIUMS REDUCED BY 20–50%

- MACHINE LIFE INCREASED BY 20–40%

- PRODUCTIVITY INCREASED BY 20–30%

- PROFIT INCREASED BY 25–60%

Figure 2-1. Benefits of predictive maintenance programs. Defines the typical benefits of a successful program.

dramatic improvement in machine life, production, operator safety, product quality, and overall profitability.

LOWER MAINTENANCE COSTS

The survey indicated that the actual costs normally associated with the maintenance operation were reduced by more than 50%. The comparison of maintenance costs included the actual labor and overhead of the maintenance department, as well as the actual materials cost of repair parts, tools, and other equipment required to maintain plant equipment. The analysis did not include lost production time, variances in direct labor, or other costs that can be directly attributed to inefficient maintenance practices.

FEWER MACHINE FAILURES

The addition of regular monitoring of the actual condition of process machinery and systems reduced the number of catastrophic, unexpected machine failures by an average of 55%. The comparison used the frequency, i.e., of number and interval, unexpected machine failures before implementing the predictive maintenance program and the failure rate during the two-year period following the addition of condition monitoring to the program. Projections of the survey results indicate that reductions of 90% can be achieved using regular monitoring of the actual machine condition.

LESS REPAIR DOWNTIME

Predictive maintenance was shown to reduce the actual time required to repair or rebuild plant equipment. The average improvement in mean-time-to-repair, MTTR, was a reduction of 60%. To determine the average improvement, actual repair times before the predictive maintenance program were compared to the actual time-to-repair after one year of operation using predictive maintenance management techniques. It was found that the regular monitoring and analysis of machine condition identified the specific failed component(s) in each machine and enabled the maintenance staff to plan each repair.

REDUCED SMALL PARTS INVENTORY

The ability to predetermine the specific repair parts, tools, and labor skills required provided the dramatic reduction in both repair time and costs. Spare parts inventories were reduced by more than 30%. Rather than carry all repair parts in inventory, the surveyed plants had sufficient lead time to order repair or replacement parts as needed. The comparison included the actual cost of spare parts and the inventory-carrying costs for each plant.

LONGER MACHINE LIFE

Prevention of catastrophic failures and early detection of incipient machine and systems problems increased the useful operating life of plant machinery by an average of 30%. The increase in machine life was a projection based on five years of operation following implementation of a predictive maintenance program. The calculation included: frequency of repairs, severity of machine damage, and actual condition of machinery following repair. A condition-based predictive maintenance program prevents serious damage to machinery and other plant systems. This reduction in damage severity increases the operating life of plant equipment.

A side benefit of predictive maintenance is the automatic ability to monitor the mean-time-between-failures, MTBF. This statistic provides the means to determine the most cost-effective time to replace machinery rather than continue to absorb high maintenance costs. The MTBF of plant equipment is reduced each time a major repair or rebuild occurs. Predictive maintenance will automatically display the reduction of MTBF over the life of the machine.

When the MTBF reaches the point that continued operation and maintenance costs exceed replacement cost, the machine should be replaced.

INCREASED PRODUCTION

In each of the surveyed plants, the availability of process systems was increased following implementation of a condition-based predictive maintenance program. The average increase in the 500 plants was 30%. The reported improvement was based strictly on machine availability and did not include improved process efficiency. However, a full predictive program that includes process parameters monitoring can also improve the operating efficiency and, therefore, the productivity of manufacturing and process plants.

One example of this type of improvement is a food manufacturing plant that had made the decision to build additional plants to meet peak demands. An analysis of existing plants, using predictive maintenance techniques, indicated that a 50% increase in production output could be achieved simply by increasing the operating efficiency of the existing production process.

IMPROVED OPERATOR SAFETY

The survey determined that advanced notice of machine-train and systems problems had reduced the potential for destructive failure, which could cause personal injury or death. The determination was based on catastrophic failures where personal injury would most likely occur. This benefit has been supported by several insurance companies that are offering reductions in premiums for plants that have a condition-based predictive maintenance program in effect.

VERIFICATION OF NEW EQUIPMENT CONDITION

Predictive maintenance techniques can be used during site acceptance testing to determine the installed condition of machinery, equipment, and plant systems. They provide the means to verify the purchased condition of new equipment before acceptance. Problems detected before acceptance can be resolved while the vendor still has reason (i.e., before the invoice is paid) to correct any deficiencies.

Many industries are now requiring that all new equipment include a reference vibration signature with the purchase. This reference signature is then compared with the baseline taken during site acceptance testing. Any abnormal deviation from the reference signature is grounds for rejection. Under this agreement, the vendor is required to correct or replace the rejected equipment.

VERIFICATION OF REPAIRS

Vibration analysis can also be used to determine whether or not repairs on existing plant machinery have corrected the identified problems and/or created additional abnormal behavior, before the system is restarted. This eliminates the need for a second outage that many times is required to correct improper or incomplete repairs.

Data acquired as part of a predictive maintenance program can be used to schedule plant outages. Many industries attempt to correct major problems or schedule preventive maintenance rebuilds during annual' maintenance outages. Predictive data can provide the information required to plan the specific repairs and other activities during the outage.

One example of this benefit was a maintenance outage scheduled to rebuild a ball mill in an aluminum foundry. Before predictive maintenance techniques were implemented in the plant, the normal outage required to completely rebuild the ball mill was three weeks, and the repair cost averaged $300,000. The addition of predictive maintenance techniques as an outage-scheduling tool reduced the outage to five days and resulted in a total savings of $200,000. The predictive maintenance data eliminated the need for many of the repairs that normally would have been included in the maintenance outage. Based on the ball mill's actual condition, these repairs were not needed. The additional ability to schedule the required repairs, gather required tools, and plan the work reduced the time required from three weeks to five days.

OVERALL PROFITABILITY

The overall benefits of predictive maintenance management have substantially improved the overall operation of both manufacturing and process plants. In all surveyed cases, the benefits derived from using condition-based management have offset the capital equipment cost required to implement the program within the first three months. Use of microprocessor-based predictive maintenance techniques has further reduced the annual operating cost of predictive maintenance methods so that any plant can achieve cost-effective implementation of this type of maintenance management program.

REASON FOR FAILURE

All of the 500 plants surveyed had successful predictive maintenance programs. There are hundreds of other companies that have
not been successful. Even though predictive maintenance is a
proven concept, many programs fail. The predominant reason is the
lack of planning and management support that are critical to a
successful program. In later chapters, the specific steps required to
implement a successful program will be discussed.

Chapter 3

Predictive Maintenance Techniques

There are a variety of technologies that can and should be used as part of a comprehensive predictive maintenance program. Since mechanical systems or machines account for the majority of plant equipment, vibration monitoring is generally the key component of most predictive maintenance programs. However, vibration monitoring cannot provide all of the information that will be required for a successful predictive maintenance program. This technique is limited to monitoring mechanical condition — not the other critical parameters required to maintain reliability and efficiency of machinery.

Therefore, a comprehensive predictive maintenance program must also include other monitoring and diagnostic techniques, such as the following: (1) thermography, (2) tribology, (3) process parameters, (4) visual inspection, and (5) other nondestructive testing techniques, including ultrasonic monitoring. This chapter will provide a description of each of these techniques, which should be included in a full capabilities predictive maintenance program for typical plants.

VIBRATION MONITORING

Vibration analysis is the dominant technique used for predictive maintenance management. Since the greatest population of typical plant equipment is mechanical, this technique has the widest application and provides the greatest benefits in a total plant program. This technique uses the noise or vibration created by mechanical equipment (and in some cases by plant systems) to determine their actual condition.

Using vibration analysis to detect machine problems is not new. During the 1960s and 1970s, the United States Navy, petrochemical industry, and nuclear electric-power-generating industry invested heavily in the development of analysis techniques based on noise or vibration that could be used to detect incipient mechanical problems in critical machinery. By the early 1980s, the instrumentation and analytical skills required for noise-based predictive maintenance were fully developed. These techniques and instrumentation had proven to be extremely reliable and accurate in detecting abnormal machine behavior. However, the capital cost of instrumentation and the expertise required to acquire and analyze noise data precluded general application of this type of predictive maintenance. As a result, only the most critical equipment in a few select industries could justify the expense required to implement a noise-based predictive maintenance program.

Recent advancements in microprocessor technology, coupled with the expertise of companies that specialize in machinery diagnostics and analysis technology, have evolved the means to provide vibration-based predictive maintenance that can be used in a cost effective manner in most manufacturing and process applications. These microprocessor-based systems simplify data acquisition, automate data management, and minimize the need for vibration experts to interpret data. Commercially available systems are capable of routine monitoring, trending, evaluating, and reporting the mechanical condition of all mechanical equipment in a typical plant. This type of program can be used to schedule maintenance on all rotating, all reciprocating, and most continuous-process mechanical equipment.

Monitoring the vibration from plant machinery can provide direct correlation between the mechanical condition and recorded vibration data of each machine in the plant. Any degradation of the mechanical condition within plant machinery can be detected using vibration-monitoring techniques. Used properly, vibration analysis can identify specific degrading machine components or the failure mode of plant machinery before serious damage occurs. Most vibration-based predictive maintenance programs rely on one or more monitoring techniques. These techniques include: broadband trending, narrowband trending, and signature analysis.

Broadband Trending

This technique acquires overall or broadband vibration readings from select points on a machine-train (Figure 3-1). These data are compared to baseline readings taken from a new machine or to vibration severity charts in order to determine the relative condition of the machine. Normally, an unfiltered broadband measurement that provides the total vibration energy between 10 and 10,000 Hertz is used for this type of analysis.

Broadband or overall RMS data are strictly gross values that represent the total vibration of the machine at the specific measurement points where the data were acquired. It does not provide any information pertaining to the individual frequency components or machine dynamics that created the measured value.

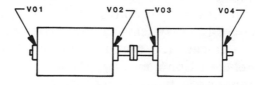

BROADBAND RMS VALUES

POINT	RMS VALUE
01	.13 IPS-RMS
02	.21 IPS-RMS
03	.05 IPS-RMS
04	.03 IPS-RMS

BROADBAND RMS

PROVIDES RELATIVE VIBRATION ENERGY
AT SELECT POINTS ON MACHINE-TRAIN

Figure 3-1. Broadband trends provide the change in overall vibration energy over a specific time period.

Narrowband Trending

Narrowband trending, like broadband, monitors the total energy for a specific bandwidth of vibration frequencies. Unlike broadband, narrowband analysis utilizes vibration frequencies that represent specific machine components or failure modes (Figure 3-2).

This method provides the means to quickly monitor the mechanical condition of critical machine components, not just the overall machine condition. This technique provides the ability to monitor the condition of gear sets, bearings, and other machine components without manual analysis of vibration signatures.

Signature Analysis

Unlike the two trending techniques, signature analysis provides visual representation of each frequency component generated by a machine-train (Figure 3-3). With training, plant staff can use vibra-

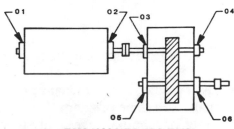

RMS VALUES (IPS-RMS)

POINT	BROADBAND	GEARSET	BRG. DEFECT
01	.12	.09	.09
02	.15	.095	.09
03	.36	.23	.10
04	.36	.25	.13
05	.30	.20	.10
06	.25	.19	.09

NARROWBAND RMS

PROVIDES RELATIVE CONDITION OF SELECT
MACHINE COMPONENTS

Figure 3-2. Narrowband trends provide the change in the vibration energy of specific machine components over a specific period of time.

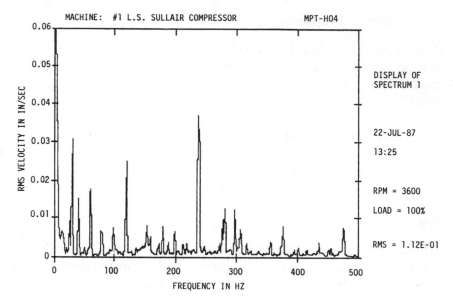

Figure 3-3. Vibration signature provides the vibration energy generated by each mechanical component within a machine.

tion signatures to determine the specific maintenance required by plant machinery.

Most vibration-based predictive maintenance programs use some form of signature analysis in their program. However, the majority of these programs rely on comparative analysis rather than full root-cause techniques. This failure limits the benefits that can be derived from this type of program.

Implementation

The capital costs for implementing a vibration-based predictive maintenance program will range from approximately $8,000 to more than $50,000. Your costs will depend on the specific techniques desired.

Training is critical for predictive maintenance programs based on vibration monitoring and analysis. Even programs that rely strictly on the simplified trending or comparison techniques require a practical knowledge of vibration theory. More advanced techniques, such as signature and root-cause failure analysis, require a working knowledge of machine dynamics and failure modes.

The chapters on establishing and maintaining a total plant predictive maintenance program will provide the practical knowledge required to implement a cost-effective vibration-based program that will provide maximum benefits.

Table 3-1 lists the plant equipment and machinery that are normally monitored using vibration analysis methods.

Table 3-1. Typical Applications for Vibration Analysis

Centrifugal

Pumps
Compressors
Blowers
Fans
Motor-Generators
Ball Mills
Chillers
Product Rolls
Mixers
Gearboxes
Centrifuges
Transmissions
Turbines
Generators
Rotary Dryers
Electric Motors, and
All rotating machinery
machinery

Reciprocating

Pumps
Compressors
Diesel Engines
Gasoline Engines, and
All reciprocating
machinery

Machine-Trains

Grinders
Boring Machines
Hobbing Mills
Machining Centers
Temper Mills
Metal Working Machines
Rolling Mills, and
Most manufacturing

Continuous Process

Continuous Casters
Hot and Cold Strip Lines
Annealing Lines
Plating Lines
Paper Machines
Can Manufacturing Lines
Pickle Lines
Printing
Dyeing and Finishing
Roofing Manufacturing Lines
Chemical Production Lines
Petroleum Production Lines
Most continuous process lines

THERMOGRAPHY

Thermography is a predictive maintenance technique that can be used to monitor the condition of plant machinery, structures, and systems. It uses instrumentation designed to monitor the emission of infrared energy (i.e., heat) to determine the operating condition. By detecting thermal anomalies—areas that are hotter or colder than they should be—an experienced surveyor can locate and define incipient problems within the plant.

Infrared technology is predicated on the fact that all objects having a temperature above absolute zero emit energy or radiation. Infrared radiation is one form of this emitted energy. Infrared "below red" emissions are the shortest wavelengths of all radiant energy and are invisible without special instrumentation. The intensity of infrared radiation from an object is a function of its surface temperature. However, temperature measurement using infrared methods is complicated because there are three sources of thermal energy that can be detected from any object: energy emitted from the object itself, energy reflected from the object, and energy transmitted by the object (Figure 3-4). Only the emitted energy is important in a predictive maintenance program. Because reflected and transmitted energies will distort raw infrared data, they must be filtered out of acquired data before a meaningful analysis can be made.

The surface of an object influences the amount of emitted or

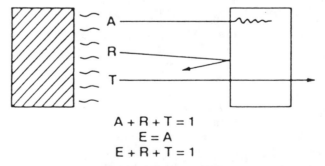

$$A + R + T = 1$$
$$E = A$$
$$E + R + T = 1$$

Figure 3-4. Energy emissions. All bodies emit energy within the infrared band. This provides the basis for infrared imaging or thermography. A = Absorbed energy. R = Reflected energy. T = Transmitted energy. E = Emitted energy.

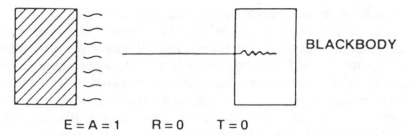

$$E = A = 1 \qquad R = 0 \qquad T = 0$$

Figure 3-5. Blackbody emissions. A perfect or blackbody absorbs all infrared energy. A = Absorbed energy. R = Reflected energy. T = Transmitted energy. E = Emitted energy.

reflected energy. A perfect emitting surface is called a "blackbody" and has an emissivity equal to 1.0 (Figure 3-5). These surfaces do not reflect. Instead, they absorb all external energy and recmit it as infrared energy. Surfaces that reflect infrared energy are called "graybodies" and have an emissivity less than 1.0 (Figure 3-6). Most plant equipment falls into this classification. Careful considerations of the actual emissivity of an object improves the accuracy of temperature measurements used for predictive maintenance.

To help users determine emissivity, tables have been developed to serve as guidelines for most common materials. However, these guidelines are not absolute emissivity values for all machines or plant equipment. Variations in surface condition, paint or other protective coatings, and many other variables can affect the actual emissivity factors for plant equipment.

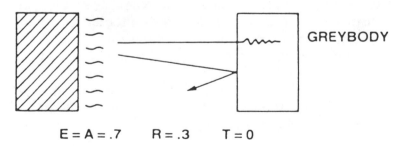

$$E = A = .7 \qquad R = .3 \qquad T = 0$$

Figure 3-6. Graybody emissions. All bodies that are not blackbodies will emit some amount of infrared energy. The emissivity of each machine must be known before implementing a thermographic program. A = Absorbed energy. R = Reflected energy. T = Transmitted energy. E = Emitted energy.

The user of thermographic techniques must consider, in addition to reflected and transmitted energy, the atmosphere between the object and the measurement instrument. Water vapor and other gases absorb infrared radiation. Airborne dust, some lighting, and other variables in the surrounding atmosphere can distort measured infrared radiation. Since the atmospheric environment is constantly changing, using thermographic techniques requires extreme care each time infrared data are acquired.

Most infrared monitoring systems or instruments provide special filters that can be used to avoid the negative effects of atmospheric attenuation of infrared data. However, the plant user must identify the specific factors that will affect the accuracy of the infrared data and apply the correct filters or other signal conditioning required to negate that specific attenuating factor or factors.

Collecting optics, radiation detectors, and some form of indicator are the basic elements of an industrial infrared instrument. The optical system collects radiant energy and focuses it upon a detector, which converts it into an electrical signal. The instrument's electronics amplify the output signal and process it into a form which can be displayed. There are three general types of infrared instruments that can be used for predictive maintenance: infrared thermometers (often called "spot radiometers"), line scanners, and imaging systems.

Infrared Thermometers

Infrared thermometers, or spot radiometers, are designed to provide the actual surface temperature at a single, relatively small point on a machine or surface. They can be used to monitor the temperature at critical points on plant machinery or equipment. This technique is typically used to monitor bearing-cap and motor-winding temperatures and to spot-check process-piping temperatures. It is limited in that the temperature represents a single point on the machine or structure. However, when used in conjunction with vibration data, point-of-use infrared data can be a valuable tool.

Line Scanners

This type of infrared instrument provides a single-dimensional scan or line of comparative radiation. Although this type of instrument provides a somewhat larger field of view (i.e., an area of machine surface), it is limited in predictive maintenance applications.

Infrared Imaging

Unlike other infrared techniques, thermal or infrared imaging provides the means to scan the infrared emissions of complete machines, process systems, or equipment in a very short time. Most of the imaging systems function much like a video camera. The user can view the thermal emission profile of a wide area by simply looking through the instrument's optics.

There are a variety of thermal-imaging instruments on the market, ranging from relatively inexpensive, black-and-white scanners to full-color, microprocessor-based systems. Many of the less expensive units are designed strictly as scanners and do not provide the capability to store and recall thermal images. The inability to store and recall previous thermal data will limit a long-term predictive maintenance program.

Implementation

Point-of-use infrared thermometers are commercially available and relatively inexpensive. The typical cost for this type of infrared instrument is less than $1,000. Infrared imaging systems will have a price range between $8,000 for a black-and-white scanner without storage capability to over $60,000 for a microprocessor-based, color-imaging system.

Training is critical with any of the imaging systems. The variables that can destroy the accuracy and repeatability of thermal data must be compensated for each time infrared data are acquired. In addition, interpretation of infrared data requires extensive training and experience.

Inclusion of thermography into a predictive maintenance program will enable you to monitor the thermal efficiency of (1) critical process systems that rely on heat transfer or retention, (2) electrical equipment, and (3) other parameters that will improve both the reliability and efficiency of plant systems. Infrared techniques can be used to detect problems in a variety of plant systems and equipment, including electrical switch gear, gearboxes, electrical substations, transmissions, circuit-breaker panels, motors, building envelopes, bearings, steam lines, and process systems that rely on heat retention or transfer. Table 3-2 lists the plant equipment that can be monitored using thermographic techniques.

TRIBOLOGY

Tribology is the general term that refers to the design and operating dynamics of the bearing-lubrication-rotor support structure of machinery. Several tribology techniques can be used for predictive maintenance: lubricating oil analysis (including spectrographic analysis), wear particle analysis (including spectrographic analysis), and ferrography.

Lubricating oil analysis is, as the name implies, an analysis technique that determines the condition of lubricating oils used in mechanical and electrical equipment. It is not a tool for directly determining the operating condition of machinery. Some forms of lubricating oil analysis will provide an accurate quantitative breakdown of individual chemical elements, both oil additives and contaminants, contained in the oil. A comparison of the amount of trace metals in successive oil samples can indicate wear patterns of oil-wetted parts in plant equipment and will provide an indication of impending machine failure.

Until recently, tribology analysis has been a relatively slow and expensive process. Analyses were conducted using traditional laboratory techniques and required extensive, skilled labor. Micropro-

Table 3-2. Equipment Monitored by Thermography

- Loss of Heat Energy
  - Facilities
  - Piping
  - Process Equipment
- Abnormal Heat Distribution
  - Machinery
  - Electrical Equipment
  - Process Equipment
- Leaks
  - Facilities
  - Piping
  - Process Equipment

cessor-based systems are now available which can automate most of the lubricating oil and spectrographic analyses, thus reducing the manual effort and cost of analysis.

The primary applications for lubricating oil analysis are: quality control, reduction of lubricating oil inventories, and determination of the most cost-effective interval for oil change. Lubricating, hydraulic, and dielectric oils can be periodically analyzed to determine their condition. The results of this analysis can be used to determine if the oil meets the lubricating requirements of the machine or application. Based on the results of the analysis, lubricants can be changed or upgraded to meet the specific operating requirements.

Detailed analysis of the chemical and physical properties of different oils used in the plant can often allow consolidation or reduction of the number and types of lubricants required to maintain plant equipment. Elimination of unnecessary duplication can reduce required inventory levels and, therefore, maintenance costs.

As a predictive maintenance tool, lubricating oil analysis can be used to schedule oil change intervals based on the actual condition of the oil. In midsize to large plants, a reduction in the number of oil changes can amount to a considerable annual reduction in maintenance costs. Relatively inexpensive sampling and testing can show when the oil in a machine has reached the point that warrants change.

The full benefit of oil analysis can only be achieved by taking frequent samples and trending the data for each machine in the plant. It can provide a wealth of information on which to base maintenance decisions. Major payback is rarely possible without a consistent program of sampling.

Lubricating Oil Analysis

Oil analysis has become an important aid to preventive maintenance. Laboratories recommend that samples be taken at scheduled intervals to determine the condition of the lubricating film that is critical to machine-train operation.

Typically, 11 tests are conducted on lube oil samples (see Figure 3-7 for an example of an oil analysis report).

VISCOSITY is one of the most important properties of a lubricating oil. The actual viscosity of oil samples is compared to an

OIL REPORT

LAB NUMBER: ZZ99011

CLIENT	
NAME: ABC MANUFACTURING ADDRESS: P.O. Box 1000 Petersburg, IN 46912	PHONE: (219) 488-3900 LOCATION: Industrial Park CONTACT: Martin Manager

UNIT	
ENGINE MAKE: Bullard ENGINE MODEL: Lathe FUEL TYPE:	MILES ON OIL: 500 Hours OIL TYPE & GRADE: Shell Tonna T-68 AMOUNT OF OIL ADDED:

COMMENTS

MARTIN: Significant increase in metals marked. Increased wear is likely attributable to 1.1% water found in the oil. Boron, silicon and sodium show this water to likely be cutting water contamination. Suggest the oil be changed and the contamination source be isolated. Further suggest another sample be taken after 30 days on fresh oil.

METAL & ADDITIVES (PPM)

ENGINE MILES	3,020	UNIT/TERMINAL AVERAGES	2,520	2,020	1,520	1,020	520	UNIVERSAL AVERAGES
SAMPLE DATE	7/8/88		6/1/88	4/25/88	2/17/88	2/17/88	1/10/88	
ALUMINUM	3	0	1	0	1	0	2	0
CHROMIUM	3	1	2	1	1	1	2	0
IRON	880	209	267	265	218	175	120	162
COPPER	127	39	45	43	32	27	51	28
LEAD	49	8	12	11	8	0	13	8
TIN	3	0	0	0	0	2	0	1
MOLYBDENUM	1	0	0	0	0	0	0	0
NICKEL	2	0	1	1	0	0	1	0
MANGANESE	13	4	6	5	4	3	6	3
SILVER	0	0	0	0	0	0	0	0
TITANIUM	0	0	0	0	0	0	0	0
VANADIUM	0	0	0	0	0	0	0	0
BORON	16	2	1	0	0	0	12	2
SILICON	7	2	4	2	1	0	7	1
SODIUM	213	63	72	68	65	69	44	44
CALCIUM	14	4	5	4	4	4	5	17
MAGNESIUM	11	7	7	6	7	6	9	4
PHOSPHORUS	678	661	648	636	655	719	651	478
ZINC	26	8	10	8	8	6	8	130
BARIUM	1	1	1	1	1	1	1	4

PROPERTIES

TEST	SUS VISCOSITY @ 100° F	SUS VISCOSITY @ 210° F	VISCOSITY INDEX	FLASHPOINT IN ° F	FUEL %	ANTIFREEZE %	WATER %	INSOLUBLES %	TBN	TAN	SULFUR	CARBON RESIDUE
VALUES SHOULD BE	49-53			390			0.05	1.1				
TEST VALUES WERE		59.7		-			1.1	0.9				

Figure 3-7. Lubrication oil analysis sample report.

unused sample to determine the thinning or thickening of the sample during use. Excessively low viscosity will reduce the oil film strength, weakening its ability to prevent metal-to-metal contact. Excessively high viscosity may impede the flow of oil to vital locations in the bearing support structure, reducing its ability to lubricate.

CONTAMINATION of oil by water or coolant can cause major problems in a lubricating system. Many of the additives now used in lubricants contain the same elements that are used in coolants. Therefore, the laboratory must have an accurate analysis of new oil for comparison.

FUEL DILUTION of oil in an engine weakens the oil film strength, sealing ability, and detergency. It may be caused by improper operation, fuel system leaks, ignition problems, improper timing, or other deficiencies. Fuel dilution is considered excessive when it reaches a level of 2.5% to 5%.

SOLIDS CONTENT is a general test. All solid materials in the oil are measured as a percentage of the sample volume or weight. The presence of solids in a lubricating system can significantly increase the wear on lubricated parts. Any unexpected rise in reported solids is cause for concern.

FUEL SOOT is an important indicator for oil used in diesel engines and is always present to some extent. A test to measure the fuel soot in diesel engine oil is important since the amount of soot indicates the fuel-burning efficiency of the engine. Most tests for fuel soot are conducted by infrared analysis.

OXIDATION of lubricating oil can result in lacquer deposits, metal corrosion, or thickening of the oil. Most lubricants contain oxidation inhibitors. However, when these additives are used up, oxidation of the oil itself begins. The degree of oxidation in an oil sample is measured by differential infrared analysis.

NITRATION results from fuel combustion in engines. The products formed are highly acidic and may leave deposits in combustion areas. Nitration will accelerate oil oxidation. Infrared analysis is used to detect and measure nitration products.

TOTAL ACID NUMBER (TAN) is a measure of the amount of acid or acidlike material in the oil sample. Because new oils contain additives that affect the TAN, it is important to compare used oil samples with new, unused oil of the same type. Regular analysis at specific intervals is important to this evaluation.

TOTAL BASE NUMBER (TBN) indicates the ability of an oil to neutralize acidity: the higher the TBN, the greater its ability to neutralize acidity. Typical causes of low TBN include: using the improper oil for an application, waiting too long between oil changes, overheating, and using high sulfur fuel.

PARTICLE COUNT tests are important in anticipating potential system or machine problems. This is especially true in hydraulic systems. The particle count analysis made as part of a normal lube oil analysis is quite different from wear particle analysis. In this test, high particle counts indicate that machinery may be wearing abnormally or that failures may occur as a result of temporarily or permanently blocked orifices. No attempt is made to determine the wear patterns, size, or other factors that would identify the failure mode within the machine.

SPECTROGRAPHIC ANALYSIS allows accurate, rapid measurement of many of the elements present in lubricating oil. These elements are generally classified as wear metals, contaminants, or additives. Some elements can be listed in more than one of these classifications.

Standard lubricating oil analysis does not attempt to determine the specific failure modes of developing machine-train problems. Therefore, additional techniques must be used as part of a comprehensive predictive maintenance program.

Wear Particle Analysis

Wear particle analysis is related to oil analysis only in that the particles to be studied are collected by drawing a sample of lubricating oil. Whereas lubricating oil analysis determines the actual condition of the oil sample, wear-particle analysis provides direct information about the wearing condition of the machine-train. Particles in the lubricant of a machine can provide significant information about the condition of the machine. This information is derived from the study of particle shape, composition, size, and quantity. Wear particle analysis is normally conducted in two stages.

The first method used for wear particle analysis is routine monitoring and trending of the solids content of the machine lubricant. In simple terms, the quantity, composition, and size of particulate matter in the lubricating oil is indicative of the mechanical condition of the machine. A normal machine will contain low levels of solids smaller than 10 microns. As the machine's condition degrades, the number and size of particulate matter will increase (see Figure 3-8).

The second wear particle method involves analysis of the particu-

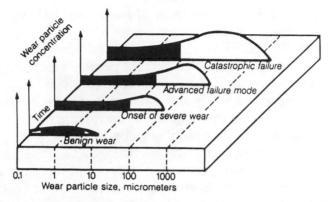

Figure 3-8. Wear particle concentration changes as wear progresses. Large-particle concentration increases as failure approaches.

late matter in each lubricating oil sample. Five basic types of wear can be identified according to the classification of particles: rubbing wear, cutting wear, rolling fatigue wear, combined rolling and sliding wear, and severe sliding wear. Only rubbing wear and early rolling fatigue mechanisms generate particles predominantly less than 15 micron in size.

RUBBING WEAR is the result of normal sliding wear in a machine. During a normal break-in of a wear surface, a unique layer is formed at the surface. As long is this layer is stable, the surface wears normally. If the layer is removed faster than it is generated, the wear rate increases and the maximum particle size increases. Excessive quantities of contaminants in a lubrication system can increase rubbing wear by more than an order of magnitude without completely removing the shear-mixed layer. Although catastrophic failure is unlikely, these machines can wear out rapidly. Impending trouble is indicated by a dramatic increase in wear particles.

CUTTING WEAR particles are generated when one surface penetrates another. These particles are produced when a misaligned or fractured hard surface produces an edge that cuts into a softer surface, or when an abrasive contaminant becomes embedded in a soft surface and cuts an opposing surface. Cutting wear particles are abnormal and are always worthy of attention. If they are only a few microns long and less than a micron wide, the cause is probably a contaminant. Increasing quantities of longer particles signal a potentially imminent component failure.

ROLLING FATIGUE is associated primarily with rolling contact bearings and may produce three distinct particle types: fatigue spall particles, spherical particles, and laminar particles. Fatigue spall particles are the actual material removed when a pit or spall opens up on a bearing surface. An increase in the quantity or size of these particles is the first indication of an abnormality. Rolling fatigue does not always generate spherical particles; these may be generated by other sources. Their presence is important because they are detectable before any actual spalling occurs. Laminar particles are very thin and are thought to be formed by the passage of a wear particle through rolling contact. They frequently have holes in them. Laminar particles may be generated throughout the life of a bearing, but at the onset of fatigue spalling the quantity increases.

COMBINED ROLLING AND SLIDING WEAR results from the moving contact of surfaces in gear systems. These larger particles (never spherical) result from tensile stresses on the gear surface, causing the fatigue cracks to spread deeper into the gear tooth before pitting. Gear fatigue cracks do not generate spheres. Scuffing of gears is caused by too high a load or speed. The excessive heat generated by this condition breaks down the lubricating film and causes adhesion of the mating gear teeth. As the wear surfaces become rougher, the wear rate increases. Once started, scuffing usually affects each gear tooth.

SEVERE SLIDING WEAR is caused by excessive loads or heat in a gear system. Under these conditions, large particles break away from the wear surfaces, causing an increase in the wear rate. If the stresses applied to the surface are increased further, a second transition point is reached: The surface breaks down and catastrophic wear ensues.

Normal spectrographic analysis of wear particles is limited to particulate contamination with a size of 10 microns or less; larger contaminants are ignored (see Figure 3-9).

Ferrography

This technique is similar to spectrography but there are two major differences. First, ferrography separates particulate contamination

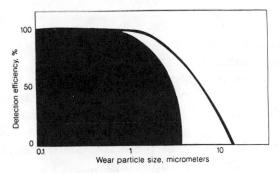

Figure 3-9. Particle detection efficiency of conventional spectrographic techniques falls off as wear-particle size approaches 10 microns.

by using a magnetic field rather than by burning a sample (as in spectrographic analysis). Because a magnetic field is used to separate contaminants, this technique is primarily limited to ferrous or other magnetic particles.

The second difference is that particulate contamination larger than 10 microns can be separated and analyzed. Normal ferrographic analysis will capture particles up to 100 microns (Figure 3-10). It provides a better representation of the total oil contamination than do spectrographic techniques.

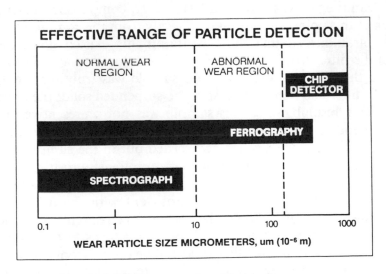

Figure 3-10. Effective range of particle detection.

Implementation

There are three major limitations when using tribology analysis in a predictive maintenance program: equipment costs, acquisition of accurate oil samples, and interpretation of data.

The capital cost of spectrographic analysis instrumention is normally too high to justify in-plant testing. Typical cost for a microprocessor-based spectrographic system is between $30,000 and $60,000. Because of this, most predictive maintenance programs rely on third-party analysis of oil samples.

Simple lubricating oil analysis by a testing laboratory will range from about $20 to $50 per sample. Standard analysis will normally include: viscosity, flash point, total insolubles, total acid number (TAN), total base number (TBN), fuel content, and water content.

More detailed analysis—using spectrographic or ferrographic techniques, including metal scans, particle size distribution, and other data—can cost well over $150 per sample.

A more serious limiting factor with any method of oil analysis is acquiring accurate samples of the lubricating oil inventory in a machine. Extreme care must be taken to obtain samples that truly represent the lubricant that passes through the machine's bearings. One recent example is an attempt to acquire oil samples from a bullgear compressor. The lubricating oil filter had a sample port on the clean (i.e., downstream) side. However, comparison of samples taken at this point and one taken directly from the compressor's oil reservoir indicated that more contaminants existed downstream from the filter than in the reservoir. Which location actually represented the oil's condition? Neither sample was truly representive. The oil filter had removed most of the suspended solids (i.e., metals and other insolubles), so this sample was not representive of the actual condition. The reservoir sample was not representive because most of the suspended solids had settled out in the sump.

Proper methods and frequency of sampling lubricating oil are critical to all predictive maintenance techniques that use lubricant samples. Sample points that are consistent with the objective of detecting large particles should be chosen. In a recirculating system, samples should be drawn as the lubricant returns to the reservoir and before any filtration. Oil should not be drawn from the bottom of a sump where large quantities of material build up over time.

Return lines are preferable to reservoirs as the sample source, but good reservoir samples can be obtained if careful, consistent practices are used. Even equipment with high levels of filtration can be effectively monitored as long as samples are drawn before oil enters the filters. Sampling techniques involve taking samples under uniform operating conditions. Samples should not be taken more than 30 minutes after the equipment has been shut down.

Sample frequency is a function of the mean-time-to-failure from the onset of an abnormal wear mode to catastrophic failure. For machines in critical service, sampling every 25 hours of operation is appropriate. However, for most industrial equipment in continuous service, monthly sampling is adequate. The exceptions to this rule of thumb are machines with extreme loads, for which weekly sampling is recommended.

Understanding the results of analysis is perhaps the most serious limiting factor. Most often, results are expressed in terms that are totally alien to plant engineers and technicians. It is therefore, difficult for them to understand the true meaning of the results in terms of oil or machine condition. A good background in quantitative and qualitative chemistry would be beneficial. At a minimum, plant staff should be required to have training in basic chemistry as well as specific instruction on interpreting tribology results.

Table 3-3 lists the plant equipment that can be monitored using tribology techniques.

PROCESS PARAMETERS

Many plants do not consider machine or system efficiency as part of the maintenance responsibility. However, machinery that is not operating within acceptable efficiency parameters severely limits the

Table 3-3. Typical Applications for Tribology

Diesel Engines
Gasoline Engines
Gas Turbines
Steam Turbines
Compressors
Hydraulic Systems

productivity of many plants. Therefore, a comprehensive predictive maintenance program should include routine monitoring of process parameters.

As an example of the importance of process parameters monitoring, consider a process pump that may be critical to plant operation. Vibration-based predictive maintenance will provide the mechanical condition of the pump, and infrared imaging will provide the condition of the electric motor and bearings. Neither technique would provide any indication of the operating efficiency of the pump. Therefore, the pump could be operating at less than 50% efficiency, and the existing predictive maintenance program would not detect the problem.

Process inefficiencies, like in the example, are often the most serious limiting factor in a plant. Their negative impact on plant productivity and profitability is often greater than the total cost of the maintenance operation. However, without regular monitoring of process parameters, many plants do not recognize this unfortunate fact.

If your program included monitoring of the suction and discharge pressures and amp load of the pump, then you could determine the operating efficiency. The brake-horsepower formula,

$$BHP = Flow\ (GPM) \times TDH \times Sp.\ Gr./3960 \times Efficiency,$$

could be used to calculate the operating efficiency of any pump in the program. By measuring the suction and discharge pressure, the total dynamic head (TDH) can be determined. The pump curve will provide the flow (in gallons per minute, or GPM) and the amp load will determine the horsepower (BHP). With these measurements, the efficiency can be calculated.

Process parameters monitoring should include all machinery and systems in the plant process that can affect its production capacity. Typical systems are the following: heat exchangers, pumps, filtration, boilers, fans, and blowers.

Inclusion of process parameters in a predictive maintenance program can be accomplished by either manual or microprocessor-based systems. Both methods will require installing instrumentation to measure the parameters that indicate the actual operating condition of plant systems. Even though most plants have installed pressure gauges, thermometers, and other instruments that should provide the information required for this type of program, many of them are no longer functioning. Therefore, including process parameters in your program may require initial capital costs to install calibrated instrumentation.

Data from the installed instrumentation can be periodically recorded using either manual logging or microprocessor-based data logging. If the latter is selected, many of the vibration-based microprocessor systems can also provide the means of acquiring process data. This should be considered when selecting the vibration-monitoring system that will be used in your program. In addition, some of the microprocessor-based predictive maintenance systems provide the ability to calculate unknown process variables. For example, they can calculate the pump efficiency used in the above example. This ability to calculate unknowns based on measured variables will enhance a total plant predictive maintenance program without increasing the manual effort required. In addition, some of these systems include nonintrusive transducers that can obtain temperatures, flow measurements, and other process data without the necessity of installing permanent instrumentation. This further reduces the initial cost of including process parameters in your program.

VISUAL INSPECTION

Regular visual inspection of the machinery and systems in a plant is a necessary part of any predictive maintenance program. In many cases, visual inspection will detect potential problems that will be missed using the other predictive maintenance techniques.

Even with the predictive techniques discussed, many potentially serious problems can remain undetected. Routine visual inspection of all critical plant systems will augment the other techniques and help insure that potential problems are detected before serious damage can occur.

Visual inspection is often considered a production department responsibility rather than a predictive maintenance technique. Many programs ignore this extremely useful tool; most of these programs fail. Visual inspection is critical to a successful program.

Most of the vibration-based predictive maintenance systems include the capability of recording visual observations as part of the routine data-acquisition process. Since the incremental costs of these visual observations are small, this technique should be incorporated in all predictive maintenance programs.

ULTRASONIC MONITORING

This predictive maintenance technique uses principles similar to those of vibration analysis. Both monitor the noise generated by plant machinery or systems to determine their actual operating condition.

Unlike vibration monitoring, ultrasonics monitor the higher frequencies (i.e., ultrasound) produced by the unique dynamics of process systems and machines. The normal monitoring range for vibration analysis is from less than 1 Hertz to 20,000 Hertz. Ultrasonic techniques monitor the frequency range between 20,000 and 100 kiloHertz.

The principal application for ultrasonic monitoring is in leak detection. The turbulent flow of liquids and gases through a restricted orifice, such as a leak, will produce a high frequency signature that can easily be identified using ultrasonic techniques. This technique is ideal for detecting leaks in valves, steam traps, piping, and other process systems.

Two types of ultrasonic systems are available that can be used for predictive maintenance: structural and airborne. Both provide fast, accurate diagnosis of abnormal operation and leaks. Airborne ultrasonic detectors can be used in either a scanning or a contact mode. As scanners, they are most often used to detect gas pressure leaks. Because these instruments are sensitive only to ultrasound, they are not limited to specific gases as are most other gas-leak detectors. In addition, they are often used to locate various forms of vacuum leaks.

In the contact mode, a metal rod acts as a waveguide. When it touches a surface, it is stimulated by the high frequencies, or ultrasound, on the opposite side of the surface. This technique is used to locate turbulent flow and/or flow restriction in process piping.

Some of the ultrasonic systems include ultrasonic transmitters that can be placed inside plant piping or vessels. In this mode, ultrasonic monitors can be used to detect areas of sonic penetration along the container's surface. This ultrasonic transmission method is useful in quick checks of tank seams, hatches, seals, caulking, gaskets, or building wall joints.

Most of the ultrasonic monitoring systems are strictly scanners that do not provide any long-term trending or storage of data. They are, in effect, a point-of-use instrument that provides an indication of the overall amplitude of noise within the bandwidth of the instrument. Therefore, the cost of this type of instrument is relatively low. The normal cost of ultrasonic instruments ranges from less than $1000 to about $8000. Used strictly for leak detection, little training is required to utilize ultrasonic techniques. The combination of low capital cost, minimum training required to use the technique, and the potential impact of leaks on plant availability provides a positive cost-benefit for including ultrasonic techniques in a total plant predictive maintenance program.

However, care should be exercised in applying this technique in your program. Many ultrasonic systems are sold as a bearing-condition monitor. Even though the natural frequencies of rolling element bearings will fall within the bandwidth of ultrasonic instruments, this is not a valid technique for determining the condition of rolling element bearings. In a typical machine, many other machine dynamics will also generate frequencies within the bandwidth covered by an ultrasonic instrument. Gear-meshing frequencies, blade-pass, and other machine components will also create energy or noise that cannot be separated from the bearing frequencies monitored by this type of instrument. The only reliable method of determining the condition of specific machine components, including bearings, is vibration analysis. The use of ultrasonics to monitor bearing condition is not recommended.

Table 3-4 lists systems and components that are normally monitored using ultrasonic techniques.

OTHER TECHNIQUES

There are numerous other nondestructive techniques that can be used to identify incipient problems in plant equipment or systems. These include: acoustic emissions, eddy-currents, magnetic particles, and residual stress. However, these techniques either do not provide a broad enough application or are too expensive to be the mainstay of a predictive maintenance program. They can be used to

Table 3-4. Typical Applications for Ultrasonic Analysis

- Detect leaks in compressed air and gas systems
- Measure corrosion and erosion in pipes and vessels
- Detect air infiltration through building envelope
- Locate leaking tubes in heat exchangers
- Detect electrical arcs and corona
- Inspect weldments and critical mechanical components
- Measure bolt pretension in critical connections

confirm failure modes identified by the predictive maintenance techniques discussed in this chapter.

If you need specific information on the techniques that are available, the American Society of Nondestructive Testing has published a complete set of handbooks that provide a comprehensive data base for most nondestructive testing techniques.

Chapter 4

A Total Plant Predictive Maintenance Program

With all of the techniques that are available for predictive maintenance, how do we select the best methods to monitor the critical machines, equipment, and systems in a plant? It would be convenient if a single system existed that would provide all of the monitoring and analysis techniques required to routinely monitor every critical piece of equipment. Unfortunately, this is not the case.

Each of the predictive techniques discussed in the preceding chapter is highly specialized. Each has a group of systems vendors who promote their technique as the *single* solution to a plant's predictive maintenance needs. The result of this specialization is that no attempt has been made by predictive maintenance systems vendors to combine all of the different techniques into a single, *total plant* system. Therefore, each plant must decide which combination of techniques and systems is required to implement its predictive maintenance program. This decision must be based on the specific goals and objectives required to meet the unique demands of the plant. Since no two plants or industries are exactly the same, there cannot be a single predictive maintenance technique or combination of techniques that is ideally suited to every plant or industry.

If a plant were to decide to use all of the available predictive maintenance techniques, a total capital cost for instrumentation and systems could easily exceed $150,000. In most cases, this fact alone would prohibit implementing a program. However, the true costs would be much higher. To implement a program that includes all of the techniques would require extensive staffing, training, and technical support. A minimum staff of at least five trained technicians and three highly trained engineers would be required to maintain this type of program. The annual cost for this type of operation would be extremely high. The actual labor and overhead costs would depend on the salaries and overhead rates of each plant, but the annual cost could easily exceed $500,000.

Because of the high capital and operating expenses, this type of program would have to save more than a million dollars each year to justify the cost. Even though this type of savings is possible in larger plants, most small and midsize plants cannot justify including all of the available techniques in their predictive maintenance programs.

How do you decide which technique or combination of techniques will provide the most cost-effective method of controlling the maintenance costs in your plant? The answer lies in the type of plant equipment to be monitored. Plants with a large population of electrical equipment, such as motors, transformers, and switch gear, should use thermographic, or infrared scanning as their primary tool. Plants with a large population of mechanical machines and systems should rely on vibration techniques.

In most cases, your plant will require a combination of two or more techniques. However, you may elect to establish one technique as the primary, in-house tool and contract with an outside source for periodic monitoring using the secondary technique or techniques. This approach would provide the benefits of the secondary techniques but without the additional costs. However, there is a downside associated with this approach. If outside contractors are used, the success or failure of your predictive maintenance program will depend on the contractor's ability to provide accurate and timely reports that you can use to schedule maintenance activities.

The demand for controlling maintenance costs has created a growing number of organizations that provide contract predictive maintenance services using one or more of the techniques discussed in the preceding chapter. Some of these organizations are excellent and would be an asset to your predictive maintenance team. Others are totally unqualified and would greatly reduce the probability of your plant's success. If you elect to use outside contractors as part of your program, great care must be exercised to select a reputable company that has a proven record of success.

THE OPTIMUM PREDICTIVE MAINTENANCE PROGRAM

Your optimum predictive maintenance program will consist of a combination of several monitoring techniques. The exact combination will depend on the specific requirements of your plant. However, most plants will need to include the following techniques for a successful program: vibration monitoring, visual inspection, process parameters, thermography, and tribology.

Vibration Monitoring

Large populations of mechanical systems account for the majority of the production capability of most plants. Vibration monitoring techniques would be the primary method recommended for implementing a total plant program in these plants.

The most common misconception associated with vibration monitoring is that the technique is limited to relatively simple, rotating equipment such as pumps, compressors, and fans. This is not true. The technique can be used for any mechanical equipment in your plant. Hundreds of successful programs have included continuous process machinery (e.g., paper machines, rolling mills, and can-manufacturing lines). In addition, assembly or production lines that have strictly linear mechanical movement can be monitored using vibration analysis techniques. Vibration monitoring is the single essential tool required for all mechanical equipment or systems.

Vibration monitoring is not the solution to all plant monitoring requirements. It cannot provide the data required to maintain electrical equipment or the operating efficiency of plant machinery or systems. Therefore, secondary methods must be used to gain this additional information.

Visual Inspection

All predictive maintenance programs should include visual inspection as one of the tools used to monitor plant systems. The cost, compared to that of other techniques that require periodic monitoring of plant equipment, is relatively small. In most cases, the visual inspection can take place as the predictive maintenance team conducts the regular data acquisition required by any of the other techniques. It, therefore, would add little or no cost to the program. Visual inspection can provide a wealth of information about the operating condition of the plant. Leaks, loose mountings, structural cracks, and a number of other failure modes that can limit the plant's performance can be detected by this simple but often neglected tool.

Most of the commercially available vibration monitoring systems build visual observation capabilities into their data acquisition instruments. Therefore, visual observations can automatically be recorded concurrently with acquisition of vibration data.

Process Parameters

A true understanding of plant conditions cannot be accomplished without knowing the operating efficiency of every machine or system in the plant. For example, how do you know the operating condition of a shell-and-tube heat exchanger without knowing the efficiency and fouling factors? The calculations required to determine these two critical factors are extremely simple, but you must first know the actual process parameters (i.e., flow, pressure, and temperature) on both the primary and secondary sides of the heat exchanger. Six simple measurements will provide the data required to calculate both the efficiency and fouling factors.

Monitoring process parameters will require, in most cases, the addition of some plant instrumentation. Few plants have working instruments that monitor all of the variables required to determine the operating condition of critical systems. However, advancements in instrumentation technology have developed nonintrusive methods of acquiring most of the required process variables without the expense of installing permanent instrumentation. For example, several techniques have been developed to monitor process flow (the most difficult process parameter to measure) without installing a Pitot or vortex-shedding flow meter. These new instruments are commercially available and, in many instances, can be read by microprocessor-based, vibration-based predictive maintenance systems.

A few of the microprocessor-based, vibration-monitoring systems provide the ability to directly acquire process data from permanently installed instruments and allow for manual entry of analog gauges. This capability provides the means to automatically acquire process parameters in conjunction with routine acquisition of vibration data. In addition, some of these systems provide the means of automatically calculating unknown process parameters such as efficiency and fouling factors. These systems record the process parameters that can be directly measured, and they automatically calculate, store, and trend the unknowns in the same manner as directly acquired parameters. This ability greatly enhances the predictive maintenance system's benefits and eliminates both the manual effort required to calculate unknowns and the potential errors that manual calculations may create.

Thermography

Implementing a full thermographic program will not always be cost effective. However, many of the vibration-based systems will permit direct acquisition of infrared data through a point-of-use scanner. This feature should be incorporated into every predictive maintenance program. The scanner can be used to acquire a number of process parameters that will augment the program. Typical applications for this technique include: bearing cap temperatures, motor-winding temperatures, spot checks of process temperatures, and spot checks of electrical equipment.

Unless the plant has a large population of electrical equipment or heat-transfer systems, the cost of implementing a full infrared scanning system is not cost-effective. For plants that have less of this type of equipment or systems, the most cost-effective method of including the benefits of full infrared scanning is to purchase periodic surveys of the plant equipment from companies that specialize in these services.

A full survey of plant equipment should be conducted at least twice each year. The frequency should be determined by the impact that these systems have on plant production. In addition to process and electrical systems, a full thermal scan of roofs and other building envelope parameters should be conducted every five years.

Tribology

Unless the plant has a large population of machinery and systems that are highly susceptible to damage as the direct result of lubricating oil contamination or has an extremely high turnover of lubricating oil inventories, the cost associated with using tribology techniques as part of a continuous predictive maintenance program is prohibitive. Even in these cases, the cost and training required to use these techniques may not be cost-effective.

The benefits of tribology can be added to your program with little incremental cost. Many of the lubricating oil suppliers can and will provide routine analysis of your lubricants for a nominal fee. In addition, there are a number of companies that provide full lubricating oil analysis on a regular schedule or as needed. Therefore, most plants can achieve the benefits of tribology without the capital or recurring costs required to perform the function in-house.

As a routine predictive maintenance tool, tribology should be limited to the simpler forms of tribology analysis, lubricating oil and spectroscopy. The data provided by these two techniques will provide all of the information required to maintain the operating condition of the plant.

Wear particle analysis should be limited to a failure analysis tool. If there is a known, chronic problem in plant machinery, this technique can provide information that will assist in the diagnostics process and the prevention of future problems. Otherwise, it is an unnecessary expense.

THE OPTIMUM PREDICTIVE MAINTENANCE SYSTEM

Based on the predictive maintenance requirements of most manufacturing and process plants, the best predictive maintenance system would use vibration analysis as the primary monitoring technique. The system should provide the ability to automate data acquisition, data management, trending, report generation, and diagnostics of incipient problems. However, the system should not be limited to this technique alone. The optimum system should include: vibration monitoring, visual inspection, process parameter monitoring, limited thermographic monitoring, and the ability to calculate process efficiencies.

In addition, the optimum system should permit direct data acquisition from any commercially available transducer. This will permit direct monitoring of any variable that may affect plant performance.

One example of this feature would be the ability to directly monitor, using a current loop tester, the electrical condition of motors. By acquiring data directly from the power cable of an electric motor and monitoring the motor's slip frequency, defects such as loose or broken rotor bars can be detected.

Selection of the specific predictive maintenance techniques and systems that will become the basis of your predictive maintenance program is the most critical step required to establish a successful program. If you believe the sales literature, each technique or system is the best and will do everything required to manage your maintenance activities. Each is touted as the most cost-effective means of controlling maintenance costs.

Few of the commercially available predictive maintenance systems will provide all of the required capabilities. Even fewer will provide the cost-effective, usable tools required to successfully implement and maintain a *total plant* predictive maintenance program. Caution *must* be exercised in your selection process. A mistake will guarantee failure of your predictive maintenance program.

Chapter 5 will provide some guidelines for selecting the right predictive maintenance system for your plant.

Chapter 5

Selecting a Predictive Maintenance System

After developing the requirements for a comprehensive predictive maintenance program, the next step is to select the hardware and software system that will most cost-effectively support your program. Since most plants will require a combination of techniques (e.g., vibration, thermography, and/or tribology), the system ideally would provide support for all of the desired techniques. Since a single system that can support all of the predictive maintenance techniques is not currently available, you must first decide on the specific techniques to use in your program. Some of the techniques may have to be eliminated to permit the use of a single predictive maintenance system, if that is preferred. Two independent systems can support the monitoring requirements of most plants.

Most plants can be cost-effectively monitored using a microprocessor-based system designed to use vibration, process parameters, visual inspection, and limited infrared temperature monitoring. Plants with large populations of heat transfer systems and electrical equipment will need to add a full thermal imaging system in order to meet the total plant requirements for a full predictive maintenance program. Plants with fewer systems requiring full infrared imaging may elect to contract this portion of the predictive maintenance program. This will eliminate the need for an additional system.

Predictive maintenance programs can be implemented without using microprocessor-based systems. However, the labor cost and staff needed to acquire and maintain the massive amount of predictive maintenance data using other types of instrumentation is prohibitive.

THE COMPLETE SYSTEM

A typical microprocessor-based system will consist of four main components: the meter or data logger (a microprocessor), host computer, software program, and transducers. Each component is important, but the total capability must be evaluated to get a system that will support a successful program.

The first step in selecting the predictive maintenance system that will be used in your plant is to develop a list of the specific features or capabilities that the system must have to support your program. As a minimum, the total system must have the following capabilities:

- User-friendly software and hardware
- Automated data acquisition
- Automated data management and trending
- Flexibility
- Reliability
- Accuracy

These capabilities are briefly discussed below, along with the training and technical support required and cost consideration.

User-Friendly Software and Hardware

The premise of predictive maintenance is that existing plant staff must be able to understand the operation of both the data logger and the software program. Since plant staff normally have little, if any, computer or microprocessor background, the system must use simple, straightforward operation of both the data-acquisition instrument and the software. Complex systems, even if they provide advanced diagnostic capabilities, may not be accepted by plant staff and, therefore, are not the appropriate basis for most long-term predictive maintenance programs.

Automated Data Acquisition

The object of using microprocessor-based systems is to remove the potential for human error, reduce manpower, and automate as much as possible the acquisition of vibration, process, and other data that will provide a viable predictive maintenance data base. Therefore, the system must be able to automatically select and set monitoring parameters without user input. The ideal system would limit user input to a single operation. However, this is not totally possible with today's technology.

Automated Data Management and Trending

The amount of data required to support a total plant predictive maintenance program is massive and will continue to increase over the life of the program. The system must be able to store, trend, and recall the data in multiple formats that will enable the user to monitor, trend, and analyze the condition of all plant equipment included in the program. The system should be able to provide long-term trend data for the life of the program. Some of the microprocessor-based systems limit trends to a maximum of 26 data sets and this will severely limit the decision-making capabilities of the predictive maintenance staff. Limiting trend data to a finite number of data sets eliminates the ability to determine the most cost-effective point to replace a machine rather than let it continue in operation.

Flexibility

The selected system must be able to support as many of the different techniques in use as possible. At a minimum, the system should be capable of obtaining, storing, and presenting data acquired from all vibration and process transducers and provide accurate interpretation of the measured values in user-friendly terms.

The minimum requirement for vibration-monitoring systems must include the ability to acquire filter broadband, select narrow-band, time traces, and high-resolution signature data using any commercially available transducer. Systems that are limited to broadband monitoring or to a single type of transducer cannot support the minimum requirements of a predictive maintenance program. The added capability of calculating unknown values based on measured inputs would greatly enhance the system capabilities. For example, neither the fouling factor nor the efficiency of a heat exchanger can be directly measured. A predictive maintenance system that can automatically calculate these values based on the measured flow, pressure, and temperature data would enable the program to automatically trend, log, and alarm deviations in these unknown yet critical parameters.

Reliability

The selected hardware and software must be proven in actual field use to ensure its reliability. The use of microprocessor-based predictive maintenance systems is still relatively new; it is important that you evaluate the field history of a system before purchase.

Ask for a users list and talk to the people who are already using the systems. This is a sure way to evaluate the strengths and weakness of a particular system before you make a capital investment.

Accuracy

Decisions on machine-train or plant system condition will be made based on the data acquired and reported by the predictive maintenance system. It must be accurate and repeatable. Errors can be input by the microprocessor and software as well as by the operators. The accuracy of commercially available predictive maintenance systems vary. While most will provide at least minimum acceptable accuracy, some are well below the acceptable level.

It will extremely difficult for the typical plant user to determine the level of accuracy of the various instruments that are available for predictive maintenance. Vendor literature and salespeople will assure the potential user that their system is the best, most accurate, and so forth. The recommended way to separate fact from fiction is to compare the various systems right in your plant. Most vendors will provide a system on consignment for periods of up to 30 days. This will provide sufficient time for your staff to evaluate each of the potential systems before purchase.

Training and Technical Support

Training and technical support are critical to the success of your predictive maintenance program. Regardless of the techniques or systems selected, your staff will have to be trained. This training should take two forms: (1) system users training, and (2) application knowledge for the specific techniques included in your program. Few, if any, of the existing staff will have the knowledge base required to implement the various predictive maintenance techniques discussed in preceding chapters. Probably none will understand the operation of the systems that are purchased to support your program.

Many of the systems manufacturers are strictly hardware and software oriented. They offer minimal training and usually no application training or technical support. Few plants can achieve minimum benefits from predictive maintenance without training and some degree of technical support. It is, therefore, imperative that the selected system or system vendors provide a comprehensive support package that includes both training and technical support.

System Cost

Cost should not be the primary deciding factor in system selection. The capabilities of the different systems vary greatly and so do the costs. It is important to compare total system capability and price before selection of your system. For example, vibration-based systems are relatively competitive in price. The general spread is less than $1,000 for a complete system. However, the capabilities of these systems are not always comparable. A system that provides minimum capability for vibration monitoring can be about the same price as one that provides full vibration-monitoring capability, process parameters, visual inspection, and point-of-use thermography.

Operating Cost

The real cost of implementing and maintaining a predictive maintenance program is not simply the cost of the initial system. Rather, it includes the annual labor and overhead costs associated with acquiring, storing, trending, and analyzing the data required to determine the operating condition of plant equipment. This is also the area where predictive maintenance systems have the greatest variance in capability. Systems that fully automate data acquisition, storage, and so forth will provide the lowest operating costs. Manual systems and many of the low-end microprocessor-based systems require substantially more personnel to accomplish the minimum objectives required by predictive maintenance. The users list can help you determine the long-term cost of the various systems. Most users will share their experience, including a general indication of labor cost.

THE MICROPROCESSOR

The data logger (a microprocessor) selected is critical to the success of the predictive maintenance program. There is a wide variety of systems on the market that range from handheld, overall-value meters to advanced analyzers that can provide an almost unlimited amount of data. The key selection parameters for a data-acquisition instrument should include: the expertise required to operate, accuracy of data, alert and alarm limits, data storage and transfer, and human resources required to meet the program demands.

Expertise Required to Operate

One of the objectives for using microprocessor-based predictive maintenance systems is to reduce the expertise required to obtain errorless, useful vibration and process data from a large population of machinery and systems within a plant. The system should not require user input to establish maximum amplitude, measurement bandwidths, or filter settings, nor should it allow free-form data input. All of these functions force the user to be a trained analyst and increase both the cost and the time required to routinely acquire data from plant equipment.

Many of the microprocessors on the market provide easy, menu-driven measurement routes that lead the user through the process of acquiring accurate data. The ideal system should require a single-key input to automatically acquire, analyze, alarm, and store all pertinent data from plant equipment. This type of system would enable an unskilled user to quickly and accurately acquire all of the requisite data.

Accuracy of Data

The microprocessor should be capable of automatically acquiring accurate, repeatable data from equipment included in the system. The elimination of user input associated with filter settings, band-widths, and other measurement parameters will greatly improve the accuracy of the data. Specific requirements that determine data accuracy vary with the type of data gathered. For example, a vibration instrument should be able to average data, reject spurious signals, auto-scale based on measured energy, and prevent aliasing.

Frequency-domain vibration analysis assumes that we monitor the rotational frequency components of a machine-train. If only a single block of data is acquired, non-repetitive or spurious data can be introduced into the data base. The microprocessor should be able to acquire multiple blocks of data, average the total, and store the averaged value. Basically, this approach enables the data-acquisition unit to automatically reject any spurious data and provide reliable data for trending and analysis.

Systems that rely on a single block of data severely limit the accuracy and repeatability of acquired data. They also limit the benefits that can be derived from the program.

The microprocessor should also have electronic circuitry that automatically checks each data set and block of data for accuracy and rejects any spurious data that may occur. Auto-rejection circuitry is available in several of the commercially available systems. Coupled with multiple-block averaging, auto-rejection circuitry assures maximum accuracy and repeatability of acquired data.

A few of the microprocessor-based systems require the user to input the maximum scale that is used to acquire data. However, this might severely limit the accuracy of the data. Setting the scale too high might prevent acquisition of factual machine data. A setting that is too low might not capture high-energy frequency components generated by the machine-train. Therefore, the microprocessor should have auto-scaling capability to ensure accurate data.

Vibration data can be distorted by high frequency components that fold-over into the lower frequencies of a machine's signature. Even though these aliased frequency components appear real, they do not exist in the machine. Low frequency components can also distort the mid-range signature of a machine in the same manner as do high frequency components. The microprocessor selected for vibration should include a full range of anti-aliasing filters to prevent the distortion of machine signatures.

The features illustrated in the vibration example also apply to non-vibration measurements. For example, pressure readings require the averaging capability to prevent spurious readings. Slight fluctuations in line or vessel pressure are normal in most plant systems. Without the averaging capability, the microprocessor cannot acquire an accurate reading of the true system pressure.

Alert and Alarm Limits

The microprocessor should include the ability to automatically alert the user to changes in machine, equipment, or system condition. Most of the predictive maintenance techniques rely on a change in operating condition to identify an incipient problem. Therefore, the system should be able to analyze data and report any change in the monitoring parameters that were established as part of the data-base development.

Predictive maintenance systems use two methods of detecting a change in the operating condition of plant equipment: static and dynamic. Static alert and alarm limits are preselected thresholds that are down-loaded into the microprocessor. If the measurement parameters exceed the preset limit, an alarm is displayed. This type of monitoring does not consider the rate of change or historical trends of a machine and, therefore, cannot anticipate when the alarm will be reached.

The second method uses dynamic limits and monitors the rate of change in the measurement parameters. This type of monitoring can detect minor deviations in the rate that a machine or system is degrading and anticipates when an alarm will be reached. The use of dynamic limits greatly enhances the automatic diagnostic capabilities of a predictive maintenance system and reduces the manual effort required to gain maximum benefits.

Data Storage

The microprocessor must be able to acquire and store large amounts of data. The memory capacity of the available predictive maintenance systems vary. At a minimum, the system must be able to store a full eight hours of data before transferring it to the host computer. The actual memory requirements depend on the type of data acquired. For example, a system used to acquire vibration data would need enough memory to store about 1,000 overall readings or 400 full signatures. Similarly, process monitoring would require a minimum of 1,000 readings to meet the minimum requirements.

Data Transfer

The data-acquisition unit is not used for long-term data storage. Therefore, it must be able to reliably transfer data into the host computer. The actual time required to transfer the microprocessor's data into the host computer is the only nonproductive time of the data-acquisition unit. It cannot be used for acquiring additional data during the data-transfer operation. Because of this, transfer time should be kept to a minimum.

Most of the available systems use a RS 232 communication protocol that allows data transfer at rates of up to 19,200 baud. At this rate, the time required to dump the full memory of a typical microprocessor can be 30 minutes or more. Some of the systems have incorporated an independent method of transferring data that eliminates the dead time altogether. These systems transfer stored data from the data logger into a battery-backed memory, bypassing the RS 232 link. Using this technique, data can be transferred at more than 350,000 baud, and the nonproductive time is reduced to a few minutes.

The microprocessor should also be able to support modem communication with remote computers. This feature enables multiple-plant operation as well as direct access to third-party diagnostic and analysis support. Data can be transferred anywhere in the world using this technique.

Not all predictive maintenance systems use a true RS 232 communications protocol or support modem communications. Those systems, if purchased, would severely limit the capabilities of your program. The various predictive maintenance techniques would add other specifications for an acceptable data acquisition unit.

THE HOST COMPUTER

The host computer provides all of the data management, storage, report generation, and analysis capabilities of the predictive maintenance program. Therefore, care must be exercised during the selection process. This is especially true if multiple technologies are to be used within the predictive maintenance program. Each predictive maintenance system has unique host-computer specifications that include hardware configuration, computer operating system, and hard-disk memory requirements. This can become a serious (if not catastrophic) problem. You may find that one system requires a special printer to provide hard copies of reports or graphic data—a printer that is not compatible with other programs. One program may operate under PC-DOS, while another requires a totally different operating system.

You should develop a complete computer specification sheet for each of the predictive maintenance systems that you plan to use. A comparison of the lists will provide—if you are lucky—a compatible computer configuration that can support each of the techniques. If this is not possible, you may have to reconsider your choice of techniques.

Computers (like plant equipment) fail. Therefore, the use of a commercially available computer is recommended. The critical considerations include: availability of repair parts and local vendor support.

Most of the individual predictive maintenance techniques will not require a dedicated computer. There is usually sufficient storage and computing capacity to handle several, if not all, of the required techniques and still leave room for other support programs such as word processing and data-base management. Use of commercially available computers gives the user the option of including these auxiliary programs in the host computer. The actual configuration of the host computer will be dependent on the specific requirements of the predictive maintenance techniques chosen. In this book, no attempt will be made to establish guidelines for selection.

THE SOFTWARE PROGRAM

The software program provided with each predictive maintenance system is the heart of a successful program. It is also the hardest to evaluate before purchase. The methodology used by each vendor of predictive maintenance systems varies greatly. Many appear to have all of the capabilities required to meet the demands of a total plant predictive maintenance program. However, upon closer inspection (usually after purchase) they are found to be lacking in some software programs.

Software is potentially the biggest limiting factor of a preventive maintenance program. Even though all vendors use one of the formal computer languages (e.g., FORTRAN, COBOL, BASIC), these are normally not interchangeable. The apparently simple task of having one computer program communicate with another can often be impossible. This lack of compatibility between various computer programs can prohibit the transfer of a predictive maintenance data base from one vendor's system into a system manufactured by another vendor. The result of this would be that, once a predictive maintenance program is started, a plant could not change to another system without losing the data already developed in the initial program.

At a minimum, the software program should provide: automatic data-base management, automatic trending, automatic report generation, and simplified diagnostics. As in the case of the microprocessor used to acquire data, the software must be user-friendly.

User-Friendly Operation

The software program should be menu-driven with clear on-line user instructions. It should prevent the user from accidentally distorting or deleting stored data. Some of the predictive maintenance systems are written in DBASE software shells. Even though these programs provide a knowledgeable user with the ability to modify or customize the structure of the program (e.g., report formats), they also provide the means to distort or destroy stored data. A single key entry can totally destroy years of stored data. Protection should be built into the program to limit the user's ability to modify or delete data and to prevent accidental data-base damage.

The program should have a clear, plain-language user's manual that provides the logic and specific instructions required to set up and use the program.

Automatic Trending

The software program should be capable of automatically storing all acquired data and updating the trends of all variables. This capability should include multiple parameters, not just a broadband or single variable. This enables the user to display trends of all variables that affect plant operations.

Most of the microprocessor based systems provide some form of automatic trending. However, most limit the number of data sets that can be included in the trend. Typically, the maximum number is 26 data sets. While 26 may sound like enough data sets to create a realistic trend for the machine or equipment condition, it might limit your ability to determine the useful life of the machinery included in your preventive maintenance program. For example, all machines have a finite useful life. As the machine ages, the Mean-Time-Between-Failures is reduced and maintenance costs increase that replacement of the machine would be more cost-effective than continued attempts to return it to service. Trends using only 26 data sets may not automatically display this decrease in MTBF. However, trends that include all data acquired from the machine will automatically display this decrease and provide the justification for machine replacement.

Automatic Report Generation

Report generation is an important part of the predictive maintenance program. Maximum flexibility in format and detail is important to program success. The system should be able to automatically generate reports at multiple levels of detail. At a minimum, the system should be able to report:

- A listing of machine-trains and other plant equipment that have exceeded, or are projected to exceed, one or more alarm limits. The report should also provide a projection of probable failure based on the historical data and last measurements.

- A listing of missed measurement points, machines overdue for monitoring, and other program management information. These reports act as reminders to ensure that the program is maintained properly.
- A listing of visual observations. Most of the microprocessor-based systems support visual observations as part of their approach to predictive maintenance. This report provides hard copies of the visual observations in addition to maintaining the information in the computer's data base.
- Equipment history reports. These reports provide long-term data on the condition of plant equipment and are valuable for analysis.

Simplified Diagnostics

Identification of specific failure modes of plant equipment requires manual analysis of data stored in the computer's memory. The software program should be able to display, modify, and compare stored data in a manner that simplifies the analysis of the actual operating condition of the equipment.

At a minimum, the software program should be able to compare data from similar machines, normalize data into compatible units, and display changes in machine parameters.

THE TRANSDUCERS

The final component of a predictive maintenance system is the set of transducers used to acquire data from plant equipment. Because we have assumed that a microprocessor-based system will be used, this discussion is limited to those sensors that can be used with this type of system.

Acquiring accurate vibration and process data necessitates several types of transducers. Therefore, the system must be capable of accepting input from as many different types of transducers as possible. Any limitation in the compatibility of the transducers can become a serious limiting factor. This should eliminate from consideration those systems that accept inputs from only a single type of transducer. Some systems are limited to a relatively small range of transducers; that will also prohibit maximum utilization of the system. Selection of the specific transducers required to monitor the mechanical condition (e.g., vibration) and process parameters (e.g., flow and pressure) deserves special consideration and will be discussed later.

Chapter 6

Establishing a Predictive Maintenance Program

The decision to establish a predictive maintenance management program is the first step toward controlling maintenance costs and improving process efficiency in your plant. What do you do next?

There are many predictive maintenance programs that can serve as models of successful implementation. Unfortunately, there are others that were aborted within the first three years because a clear set of goals and objectives was not established before the program was implemented. Implementing a total plant predictive maintenance program is not inexpensive. In addition to the initial capital costs of instrumentation and systems, there are substantial annual labor costs required to maintain the program.

To be successful, a predictive maintenance program must be able to quantify the benefits generated by the program. This can be achieved if the program is properly established, uses the proper predictive maintenance techniques, and has measurable benefits. The amount of effort expended to initially establish the plant program is directly proportional to its success or failure.

Proper implementation of a predictive maintenance program must include the components listed in Table 6-1. These are discussed below.

GOALS, OBJECTIVES, AND BENEFITS

Constructive actions issue from a well established purpose. It is important that the goals and objectives of a predictive maintenance program be fully developed and adopted by the personnel who

Table 6-1. A Successful Program Must Have:

- Clearly defined objectives and goals
- Measurable benefits
- Management support
- Dedicated, accountable personnel
- Efficient data-collection and analysis procedures
- Adequate record-keeping & information organization procedures
- Communications capability
- Evaluation procedures

perform the program and by the upper management of the plant. A predictive maintenance program should not be an excuse to buy sophisticated, expensive equipment. Neither should its purpose be to keep a number of people busy measuring and reviewing data from the various machines, equipment, and systems within the plant.

The real purpose of predictive maintenance is to minimize unscheduled equipment failures, maintenance costs, and lost production. It is also intended to improve the production efficiency and product quality in your plant.

This is accomplished by regular monitoring of the mechanical condition, machine and process efficiencies, and other parameters that define the operating condition of the plant. Using the data acquired from critical plant equipment, incipient problems are identified and corrective actions taken to improve the reliability, availability, and productivity of the plant.

Specific goals and objectives will vary from plant to plant. Before they can be developed for your plant, you must determine the existing maintenance costs and other parameters in order to establish a reference, or baseline, data set. Since most plants do not track the true cost of maintenance, this may be the most difficult part of establishing a predictive maintenance program. As a minimum, your baseline data set should include the labor, overhead, overtime premiums, and other payroll costs of the maintenance department. It should also include all maintenance-related contract services (excluding janitorial) and the total cost of spare parts inventories.

The baseline should also include the percentage of unscheduled maintenance repairs vs. scheduled ones, actual repair costs on critical plant equipment, and the annual availability of the plant. This baseline should include the incremental cost for production created by catastrophic machine failures and other parameters. If these costs are available or can be obtained, they will be of great help in establishing a valid baseline.

The long-term objectives of a predictive maintenance program are to:

- Eliminate unnecessary maintenance
- Reduce rework costs
- Reduce lost production caused by failures
- Reduce repair parts inventory
- Increase process efficiency
- Improve product quality
- Extend the operating life of plant systems
- Increase production capacity
- Reduce overall maintenance costs
- Increase overall profits

Eliminate Unnecessary Maintenance

As discussed in earlier chapters, one reason for implementing a predictive maintenance program is to reduce maintenance costs by eliminating unnecessary repairs. Maintenance management programs that rely on time-based, preventive maintenance programs schedule machine repairs based on the elapsed time in service. Predictive maintenance will provide the means to schedule repairs on an as-needed basis. This reduction in unnecessary repairs will reduce maintenance costs. A well-designed and properly maintained predictive maintenance program should reduce the percentage of unnecessary repairs by 90%.

Reduce Rework Costs

It is one of Murphy's Laws that we often do more damage to a machine during repair than we would if we allowed the machine to run without repair. In other cases, we repair part of the problem and within a few days must again remove the machine from service to repair other problems that were missed during the first repair. Predictive maintenance provides the means to identify the specific problem or problems within a machine and permit complete repair the first time. In addition, it provides the means to identify new problems created by the repair *before* returning the machine to service. Rework costs created by improper or incomplete repair can be totally eliminated by a well-conceived and properly implemented predictive maintenance program.

Reduce Lost Production Caused by Failures

Loss of critical production machinery during peak production periods is one of the major reasons for implementing a predictive maintenance program. The ability to detect and identify an incipient problem *before* catastrophic failure will reduce the number of unexpected failures in your plant.

The reduction in the number of unexpected failures will depend on your existing record. If your plant has more than 30% of its repairs conducted in the reactive, after-failure mode, predictive maintenance will greatly improve this part of your maintenance cost. In a plant using predictive maintenance, the number of unexpected failures should be less than 10% of all repairs.

Reduce Repair Parts Inventory

The average reduction in spare parts inventory should be 25% to 30%. The actual percentage will depend on your existing program.

Increase Process Efficiency

Predictive maintenance programs that include process efficiency monitoring should be able to improve the overall efficiency of the plant by at least 30%. Improvements of more than 50% have been achieved by comprehensive programs. You may be surprised by how inefficient your plant systems really are.

One example of the impact of process inefficiency can be illustrated by a snack food manufacturer. This company had decided to build new plants to meet the consumer demands during four peak periods each year. As an afterthought, they asked a plant performance consultant to measure the operating efficiency of their existing plants. The analysis confirmed that the production capacity of their existing plants could be doubled with minor corrections of existing process inefficiencies. The results of the analysis eliminated the need to build new plants.

Improve Product Quality

Product quality improvement is difficult to quantify. In most programs, this is a side benefit that is not used to justify the program's cost. However, correction of mechanical problems, inefficiency of process systems, and all of the other benefits of predictive maintenance will automatically improve product quality.

Extend Operating Life of Plant Systems

Detection and correction of plant problems *before* they can cause serious damage to plant systems will not only reduce maintenance costs but also extend the useful operating life of your plant. Extensions of five to ten years are within reason.

Increase Production Capacity

Production capacity increases of 10% to 30% or more are possible when a comprehensive predictive maintenance program is implemented. As in the previous cases, the actual amount will depend on the existing maintenance management program.

Reduce Overall Maintenance Costs

The combination of the stated savings will reduce overall maintenance costs in your plant. Typical reductions of 25% to 50% are commonplace.

Increase Overall Profits

Profits can be increased by 50%. The impact of reduced maintenance costs on the bottom-line profit of any company can be great enough to justify the program cost.

However, just stating these objectives will not make them happen nor will it provide the means of measuring the success of the program. You must establish specific objectives (e.g., reduce unscheduled maintenance by 20% or increase production capacity by 15%), and then define the methods that will be used to accomplish each objective as well as those used to measure the actual results.

MANAGEMENT SUPPORT

Implementing a predictive maintenance program requires an investment in both capital equipment and personnel. If a program is to get started and survive long enough to accomplish its intended goals, management must be willing to commit the necessary resources. They must also insist on the adoption of vital record-keeping and information-exchange procedures that are critical to program success and are usually outside the control of the maintenance department.

In most aborted programs, management had committed the initial investment for capital equipment but had not invested the resources required for training, consulting support, and in-house staff—all of which are essential for success. A number of programs were aborted between 18 and 24 months following implementation. These programs were not aborted because the program failed to achieve the desired results. They failed because upper management did not clearly understand how the program worked or the potential long-term benefits of the program.

During the first 12 months, most predictive maintenance programs identify numerous problems in plant machinery and systems. Therefore, the reports and recommendations for corrective actions generated by the predictive maintenance group are highly visible. After the initial 12 to 18 months, most of the serious plant problems have been resolved, and the reports begin to show little need for corrective actions. Without a clear understanding of this normal cycle and the means of quantifying the achievements of the predictive maintenance program, upper management often concludes that the program is not providing sufficient benefits to justify the continued investment of human resources.

DEDICATED AND ACCOUNTABLE PERSONNEL

All successful programs are built around a full-time predictive maintenance team. Some of these teams cover multiple plants and some monitor only one. However, every successful program has this dedicated team who can concentrate their full attention on achieving the objectives established for the program.

Even though a few successful programs have been structured around part-time personnel, this approach cannot be recommended. All too often, the part-time personnel will not or cannot maintain the frequency of monitoring and analysis that is critical to a successful program.

The accountability expected of the predictive maintenance group is another factor that is critical to program effectiveness. If measures of program effectiveness are not established, neither management nor program personnel can determine if the program's potential is being achieved.

EFFICIENT DATA-COLLECTION AND ANALYSIS PROCEDURES

Efficient procedures can be established if adequate instrumentation is available and the monitoring tasks are structured to emphasize program goals. A well-planned program should not be structured so that all machines and equipment in the plant receive the same scrutiny.

Typically, predictive maintenance programs monitor from 50 to 500 machine-trains in a given plant. Some of the machine-trains are obviously more critical to the continued, efficient operation of the plant than others. The predictive maintenance program should be set up with this in mind and concentrate the program's efforts in the areas that will provide maximum results.

The use of microprocessor-and-personal-computer-based predictive maintenance systems will greatly improve the data-collection and data-management functions required for a successful program. They can also provide efficient data analysis. The methods, schedule, and other parameters of data acquisition, analysis, and report generation must be included in the program definition.

A VIABLE DATA BASE

The methods and systems that you choose for your program and the initial program development will, to a great extent, determine the success or failure of predictive maintenance in your plant.

Proper implementation of a predictive maintenance program is not easy. It will require a great deal of thought and (perhaps for the first time) a complete understanding of the operation of the various systems and machinery in your plant.

The initial data-base development required to successfully implement a predictive maintenance program may involve many staff-months of effort. The lack of sufficient labor to properly establish a predictive data base often results in either a poor or an incomplete one. In some cases, the program is discontinued as a result of staff limitations. If the extensive labor required to establish a data base is not available in-house, there are consultants available who will provide the knowledge and labor required to accomplish this task. The ideal situation would be to have the predictive maintenance systems vendor establish a viable data base as part of the initial capital equipment purchase. This service is offered by a few systems vendors.

There are a variety of technologies and predictive maintenance systems that can be beneficial. How do you decide which method and system to use? A vibration-based predictive maintenance program is the most difficult to properly establish and will require much more effort than any of the other techniques. It will also provide the most return on investment. Too many of the vibration-based programs fail to use the full capability of this predictive maintenance tool. They ignore the automatic diagnostic power that is built into most of the microprocessor-based systems and rely instead on manual interpretation of all acquired data.

The first step is to determine the types of plant equipment and systems that are to be included in your program. A plant survey of your process equipment should be developed that lists every critical component within the plant and its impact on both production capacity and maintenance costs. A plant process layout is invaluable during this phase of program development. It is very easy to omit critical machines or components during the audit. Therefore, care should be taken to ensure that all components that can limit production capacity are included in your list.

The listing of plant equipment should be ordered into the following classes depending on their impact on production capacity or maintenance cost: essential, critical, serious, and others.

Class I, or essential machinery or equipment, must be on-line for continued plant operation. Loss of any one of these components will result in a plant outage and total loss of production. Any plant equipment that has excessive repair costs or repair-parts lead-time should also be included in the essential classification.

Class II, or critical machinery, would severely limit production capacity if disabled. As a rule of thumb, loss of critical machinery would reduce production capacity by 30% or more. Also included in the critical classification are machines or systems with chronic maintenance histories or that have high repair or replacement costs.

Class III, or serious machinery, includes major plant equipment that does not have a dramatic impact on production but that contributes to maintenance costs. An example of the serious classification would be a redundant system. Since the in-line spare could maintain production, loss of one component would not affect production. However, the failure would have a direct impact on maintenance costs.

Class IV machinery would include other plant equipment having a history of impacting either production or maintenance costs. All equipment in this classification must be evaluated to determine whether routine monitoring is cost-effective. In some cases, replacement costs are lower than the annual costs required to monitor machinery in this classification.

The completed list should include every machine, system, or other plant equipment whose incapacity could have a serious impact on the availability and process efficiency of your plant.

The next step is to determine the best method or technique for cost-effectively monitoring the operating condition of each item on the list. To select the best methods for regular monitoring, you should consider the dynamics of operation and normal failure modes of each machine or system to be included in the program. A clear understanding of the operating characteristics and failure modes will identify which predictive maintenance method should be used.

Most predictive maintenance programs should use vibration monitoring as the principal technique. The inclusion of visual inspection, process parameters, ultrasonics, and limited thermographic techniques should also be added to the in-house program. The initial cost of systems and advanced training required by full thermographic and tribology techniques prohibit their inclusion into most in-house programs. Plants that require these techniques normally rely on outside contractors to provide the instrumentation and expertise need for these monitoring and diagnostic techniques.

Because of the almost unlimited number and variety of machinery and systems used for predictive maintenance, it is impossible to cover each of them in this book. However, Chapter 7 provides a cross section that illustrates the processes used to identify the monitoring parameters for plant equipment.

Chapter 7

Monitoring Parameters
for Plant Machinery

One of the critical steps required for a successful predictive maintenance management program is the selection of the most cost-effective methods for monitoring the machine, equipment, and system parameters that provide early warning of incipient problems. To accomplish this task, you must have a complete understanding of how the machines, plant equipment, and systems operate. These can be divided into several classifications: mechanical, electrical, heat transfer, hydraulic, and others. Each group has unique characteristics and operating parameters that are indicative of their condition and efficiency. They must be considered before selecting the predictive maintenance technique.

There will be some overlap or combinations of classifications. For example, an electric circuit breaker is classified primarily as an electrical system, although it also has some mechanical characteristics. The normal predictive test for circuit breakers is to rack-out the breaker and run continuity tests on the unit. Opening and closing the contacts of a circuit breaker can itself cause problems or failure of the unit. An alternative to this test procedure is to use an infrared scanner to view the thermal image of the circuit breaker in operation. Any abnormal characteristics will be clearly displayed as hot spots. This technique provides the means to detect incipient problems without opening the circuit breaker.

Plant equipment that combines several classifications may require more than one predictive maintenance monitoring technique.

MECHANICAL EQUIPMENT

Mechanical equipment includes all rotating, reciprocating, and other plant equipment or systems that have moving components. This classification includes pumps, fans, compressors, motor-generators, conveyors, paper machines, and many other continuous-process machines. The primary method for the routine monitoring of mechanical condition is vibration. A properly implemented predictive maintenance program regularly monitors the vibration at specific locations on each mechanical machine-train in the program. By definition, a machine-train consists of a primary driver or drivers (e.g., electric motor, turbine), all intermediate drives (e.g., couplings, belts, gearbox), and all driven machine components (e.g., fans, pumps, continuous processes).

Each component within a machine-train will generate specific mechanical or dynamic forces during operation of the machine. Each of these forces generates, in turn, specific vibration frequencies that uniquely identify the machine component. For example, a gear set generates a unique set of vibration frequencies that identify the actual, normal meshing of the gears. Any degradation of the gear set changes the amplitude and spacing of the unique vibration frequencies generated by the gear set. Because the individual components of a machine-train are mechanically linked, the vibration frequencies generated by each individual machine component are also transmitted throughout the machine-train. Monitoring the vibration frequencies at specific points throughout the machine-train can be used to isolate and identify the specific machine component that is degrading.

To achieve maximum benefit and diagnostic power from your predictive maintenance program, you must monitor and evaluate the total machine-train. Many programs have been severely limited by monitoring only some machine-train components. This approach limits the ability for early detection of incipient problems. For example, how do you determine that the shafts between a driver and a driven machine component are misaligned if you cannot compare the unique vibration-frequency components on both sides of the coupling? Initially, this knowledge of the operating dynamics of plant machinery is mandatory for the successful establishment of the vibration-monitoring data base, monitoring frequency, alert/

alarm limits, and analysis parameters. Later, it can provide the basis for analyzing the vibration data to determine the degrading machine component(s), severity, and root-cause of incipient machine problems.

Vibration-based predictive maintenance can be applied to an almost infinite variety of process and manufacturing machinery. However, this section will be limited to machinery that is common to most processes. It is intended to provide guidelines for the information required to implement this type of monitoring on all mechanical plant equipment.

All mechanical plant equipment can be broken down into four classifications: (1) constant speed, constant load; (2) constant speed, variable load; (3) variable speed, constant load; and (4) variable speed, variable load. The vibration-monitoring system that you select must be able to handle effectively all of these combinations of machine operations. Why is this important? Both speed and load affect the location and amplitude of the unique vibration components generated by the mechanical forces or motion within the machine-train.

The location, or frequency, of individual vibration components maintains a fixed relationship to the actual running speed of the specific shaft that generates the force. As the shaft speed changes, so will the location (frequency) of the individual vibration components generated by that shaft. For example, the gear-meshing frequency component of a gear with 10 teeth, mounted on a shaft turning at 20 Hertz, is located at 200 Hertz. If the shaft speed changes to 40 Hertz, the gear-meshing frequency will moves to 400 Hertz.

Load changes will not cause the location of individual vibration-frequency components to change but will affect the amplitude, or energy, of each component. Changes in machine load will either amplify or dampen the energy of individual vibration components. The variation of vibration energy at 100% load cannot be directly compared to the same machine operating at 50% load. Therefore, your vibration-based predictive maintenance program must compensate for load variations.

Machine-Train Components

To establish and utilize vibration-based predictive maintenance, a complete knowledge of all machine components and how they interact within the machine-train is absolutely necessary. Every phase of a predictive maintenance program, from implementation through root-cause failure analysis, is driven by the dynamics and the resulting vibration characteristics of each machine-train. All rotating, reciprocating, and continuous-process machinery have common components, characteristics, and failure modes. Yet each machine also has totally unique operating dynamics and failure modes. In the following sections, we will discuss both the common and unique characteristics of typical machine-trains found in most manufacturing and process plants.

All plant machinery have the following common components and characteristics that affect their vibration signature:

- Bearings
- Gears and gearboxes
- Running speeds
- Critical speeds
- Mode shape
- Resonance
- Preloads and induced loads
- Process variables
- Blades and vanes
- Belt drives

Bearings

In modern machinery operating at relatively high speeds and loads, the proper selection and design of the bearings are the primary limiting factors in the operating life of the machine-train. The first indication of machinery problems often develops in the vibration signature of the machine's bearings. However, bearings are typically not the cause of machine-train problems. Because they are, by design, the weakest link in most machinery, bearings are usually the first point of machine failure. As a result, it is beneficial to have a good understanding of bearing design and operating dynamics.

Bearings can be divided into two classifications: rolling-element and sleeve (Figure 7-1). The two classifications have unique operating characteristics and failure modes that can be monitored using vibration analysis techniques.

Rolling-Element Bearings.   Over the past several decades, rolling-element (i.e., ball or roller) bearings have been used for most high-

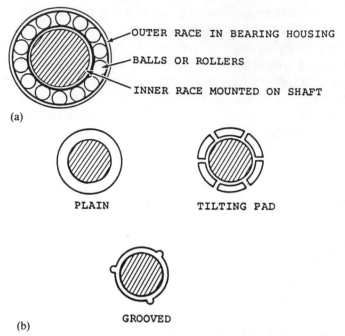

Figure 7-1. Bearing types: (a) Antifriction or rolling element bearings, (b) Sleeve or babbitt bearings.

speed applications and in most smaller process machinery. The primary components of a rolling-element bearing are the outer race, inner race, cage, and rolling elements. General bearing behavior is determined by the interaction between these various elements. The rolling-element-to-race contacts are the heaviest loaded; hence, most fatigue failures involve this interaction. The rolling-element-to-cage and race-to-cage contacts are generally dynamic in nature since they constitute a number of very high-speed, short-time-frame collisions.

There are many factors that affect normal bearing life (e.g., lubrication, design loads, bearing design), but this discussion is limited to the mechanical factors that can be used to predict bearing failure.

Defective rolling-element bearings generate vibration frequencies at the rotational speeds of each bearing component. Each of the frequencies can be calculated and routinely monitored using vibration-analysis techniques. Rotational frequencies are related to the

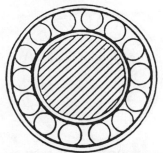

BALL SPIN FREQUENCY (BSF) IS THE ROTATING SPEED OF THE BALL OR ROLLER

BSF	=	$D/d \cdot FN \cdot [1 - (d/D)^2 \cdot \cos^2 \beta]$
D	=	PITCH DIAMETER
d	=	BALL/ROLLER DIAMETER
FN	=	RELATIVE SPEED OF INNER & OUTER RACE
β	=	CONTACT ANGLE OF BEARING

BALL DEFECT WILL APPEAR AT 1x BSF

$BSF = 0.5 N \cdot D/d [1 - (d/D)^2]$

BALL DEFECT WILL APPEAR AT 2x BSF

Figure 7-2. Ball spin defect frequency (BSF) will be generated by defects or spalls on a roller or ball.

motion of the rolling elements, cage, and races. They include: ball or roller spin, cage rotation, and ball or roller passing frequencies.

The ball or roller spin frequency (BSF) is generated by the rotation of each ball or roller around its own centerline and can be calculated as

$$BSF = 0.5N \times (D/d) \times [1 - (d/D)^2],$$

where: N = shaft speed, in Hertz or revolutions per second,
D = pitch diameter of bearing, in inches,
d = diameter of balls or roller, in inches.

Since a ball or roller defect contacts both the inner and outer race each time the ball or roller completes one complete revolution, the

FUNDAMENTAL TRAIN FREQUENCY (FTF) ROTATING FREQUENCY OF BEARING CAGE IN RELATION TO INNER RACE

FTF	=	0.5 N [1 - d/D]
N	=	SHAFT 1x SPEED
d	=	BALL/ROLLER DIAMETER
D	=	PITCH DIAMETER OF BEARING

TYPICALLY FTF IS APPROXIMATELY .4x ROTATING SPEED OF SHAFT

CAGE DEFECT WILL APPEAR AT 1x FTF

Figure 7-3. Cage defect frequency or fundamental train frequency (FTF) will be generated by defects in the bearing cage.

ball spin defect frequency will be at two times the BSF, or rotational frequency.

The cage rotation frequency, or fundamental train frequency (FTF), can be calculated as

$$FTF = 0.5N \times [1 - (d/D)].$$

A defect in the outer race of the bearing can be calculated using the ball pass frequency–outer race (BPFO) formula,

$$BPFO = 0.5Nn \times [1 - (d/D)],$$

where: n = number of balls or rollers.

Inner-race defect frequency, or ball frequency pass–inner race (BPFI), can be calculated as

$$\text{BPFI} = 0.5Nn \times [1 + (d/D)].$$

Many bearing manufacturers have simplified the calculation of bearing defect frequencies by providing a bearing reference guide. This guide provides a constant (i.e., a number having a fixed value) for each defect frequency of each bearing manufactured by the vendor. This constant is multiplied by the actual running speed of the machine's shaft to obtain the unique defect frequencies.

Bearing rotational and defect frequencies may be generated as the result of actual bearing defects or by machine- or process-induced loads. Imbalance, misalignment, and abnormal loads amplify the

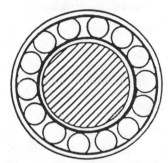

BALL PASS OUTER RACE (BPFO) ROTATING FREQUENCY OF BALL/ROLLERS TO OUTER RACE.

BPFO = 0.5 Nn (1 - d/D) COS β

N = RELATIVE SPEED OF INNER & OUTER RACE

n = NUMBER OF BALLS OR ROLLERS

d = BALL/ROLLER DIAMETER

D = PITCH DIAMETER

β = CONTACT ANGLE

OUTER RACE DEFECT WILL APPEAR AT 1x BPFO

Figure 7-4. Outer race defect frequency or ball-pass frequency outer race (BPFO) will be generated by defects in the outer race.

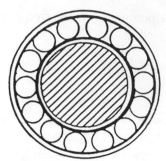

BALL PASS INNER RACE FREQUENCY (BPFI) ROTATING
FREQUENCY OF BALL/ROLLER TO INNER RACE.

BPFI = 0.5 Nn (1 + d/D) COS β

N = RELATIVE SPEED OF INNER & OUTER RACE

n = NUMBER OF BALLS OR ROLLERS

d = BALL/ROLLER DIAMETER

D = PITCH DIAMETER

β = CONTACT ANGLE

INNER RACE DEFECT WILL APPEAR AT 1x BPFI

Figure 7-5. Inner race defect frequency or ball-pass frequency inner race (BPFI) will be generated by defects in the inner race.

specific bearing frequencies that must absorb the load. For example, excessive bearing side load created by too much belt tension will amplify the ball spin frequency and both ball pass frequencies. This is the direct result of the abnormal load created by the belt tension. Misalignment of the same belt drive will also amplify the cage frequency (FTF).

The unique vibration frequencies that define incipient bearing problems can be easily identified using narrowband monitoring techniques. The fundamental train frequency (FTF) or cage defect always occurs at about 40% of running speed (Figure 7-6). A narrowband established to monitor the energy in a frequency band from 30% to 50% of running speed will automatically detect any abnormal change in the condition of the bearing's cage. Of the remaining three bearing defect frequencies, the ball spin frequency (BSF) is always the lowest frequency. The ball pass frequency–

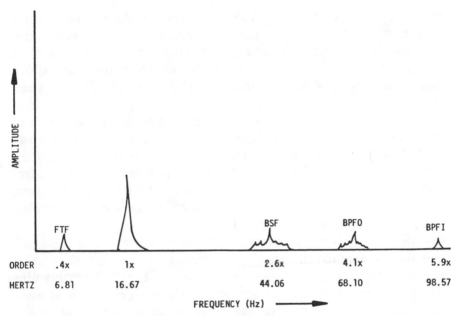

Figure 7-6. A typical rolling element bearing signature will show all of the rotational frequencies of the bearing.

(BPFI) inner race is always the highest. A single narrowband can be established to monitor these bearing defect frequencies. The narrowband should be established with the lower limit set about 10% below the actual ball spin frequency (BSF) to allow for slight variations in running speed. The upper limit should be set about 10% higher than the actual inner race defect frequency (BPFI). By using these narrowband monitoring techniques, the microprocessor-based data-acquisition unit can automatically detect any abnormal change in the bearing's condition.

The narrowband monitoring the three higher defect frequencies cannot identify the specific defect (i.e., inner race, outer race, or ball) but reports the operating condition of the bearing. If you want to know which bearing component is degrading, the full signature must be manually evaluated to detect the specific defect frequency that is in alarm.

Sleeve Bearings. Sleeve, or fluid-film bearings can be divided into several subclasses: plain, grooved, partial arc, and tilting pad. With the exception of the tilting-pad bearing, they do not generate a unique rotational frequency that would identify normal operation. Because the tilting-pad bearing has moving parts (i.e., the pads), it will generate low-level vibration components at a pad-passing frequency that is equal to the number of pads multiplied by the shaft running speed.

The tilting-pad bearing is designed to form a uniform thin film of lubricant between the bearing's babbitt surface and the rotating shaft. In normal operation, the shaft is centered in this thin lubricating film and creates neither a dynamic force nor a vibration frequency component that uniquely identifies the bearing. However, abnormal behavior of the lubricating film can be clearly identified using vibration analysis techniques.

If the lubricating film becomes eccentric, the vibration signature will show a marked increase in low-frequency (i.e., less than shaft running speed) energy. Initial breakdown of the uniform oil film will be indicated by an increase of frequency components at even fractions (e.g., ¼, ⅜, ½) of the running speed. These vibration components are created by the eccentric rotation of the machine's shaft (Figure 7-7). As the condition degrades, the fractional vibration components will consolidate between about 40% and 48% of the shaft's true running speed.

If the shaft breaks through the lubricating film, a mechanical rub becomes evident in the vibration signature. This rub is shown as a very low frequency signature (i.e., between 1.0 and 2.0 Hertz) and has low-amplitude components at about 25% and 40% of true running speed. The rub signature has the appearance of a ski jump and, in most cases, will generate high amplitudes.

Only a limited number of the microprocessor-based vibration-monitoring systems are capable of detecting vibration frequencies below 10 Hertz (600 RPM). All others cannot be used to detect the mechanical rubs that often occur in machinery.

Monitoring the mechanical condition of bearings should rely primarily on vibration-analysis techniques. However, periodic lubricating oil analysis using either spectrographic or wear particle methods can provide additional data on the actual operating condition. The added costs associated with these techniques do not justify their use unless a specific problem has been identified by the vibration-monitoring program.

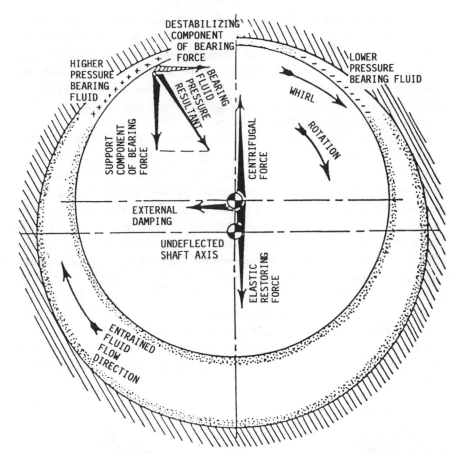

Figure 7-7. Oil whip/oil whirl are the normal failure modes associated with sleeve or babbitt bearings.

Bearing temperatures can be added to the monitoring program using point-of-use infrared sensors that can input temperature data directly into the vibration-monitoring instrumentation. The incremental cost is very low and the added temperature data will assist in the early identification of incipient problems.

Gears and Gearboxes

Many machine-trains use gear drive assemblies to connect the driver to various driven machine components. Gears and gearboxes have unique vibration signatures that identify both normal and abnormal operation. Characterization of the vibration signatures of a gearbox is difficult to acquire, but it is an invaluable tool for diagnosing machine-train problems. The difficulty is due to two factors: (1) It is difficult to mount vibration transducers close to the individual gears within a gearbox; and (2) the number of vibration sources in a multi gear drive results in a complex assortment of gear mesh, modulation, and running-speed frequencies.

Severe gearbox vibration is usually due to resonance between the system's natural frequency and the shaft speeds. This resonant excitation arises from, and is proportional to, gear inaccuracies that cause small periodic fluctuations in the pitch-line velocity. Complex machines usually have many resonance frequencies within their operating-speed range. At resonance, these cyclic excitations may cause large vibration amplitudes and stresses. Basically, the forcing torque arising from gear inaccuracies is small. However, under resonant conditions, torsional amplitude growth is restrained only by damping in the mode of vibration. In typical gearboxes, this damping is often small and permits the gear-excited torque to generate large vibration amplitudes under resonant conditions.

One other important fact about gear sets is that all of them have a designed preload and create an induced load in normal operation.

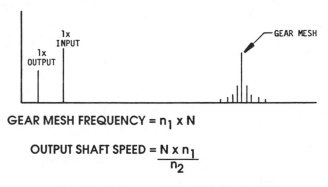

GEAR MESH FREQUENCY = n_1 x N

OUTPUT SHAFT SPEED = $\dfrac{N \times n_1}{n_2}$

n_1 = NUMBER OF TEETH ON DRIVE (INPUT) GEAR

n_2 = NUMBER OF TEETH ON DRIVEN (OUTPUT) GEAR

N = ROTATING SPEED OF DRIVE (INPUT) SHAFT

Figure 7-8. Normal gear profile or signature.

The direction and force of these loads vary depending on the type of gears and gearbox design.

The following description of typical gears will provide some insight into the normal operating dynamics of each gear type. To implement a vibration-based predictive maintenance program for gears and gearboxes, a basic knowledge of the dynamic forces that they generate is very important. At a minimum, the following forces and their corresponding vibration components should be identified: gear mesh, gear excitation, and backlash.

Gear Mesh.   This is the frequency most commonly associated with gears and is equal to the number of teeth on the gear multiplied by the actual running speed of its shaft. A typical gearbox has multiple gears and, therefore, multiple gear-meshing frequencies. A normal gear-mesh signature has a low-amplitude gear-mesh frequency with a series of symmetrical sidebands, spaced at the exact running speed of the shaft, on each side of the mesh components (Figure 7-9). The spacing and amplitude of these sidebands will be exactly symmetrical if the gearbox is operating normally. Any deviation in the symmetry of the gear-mesh signature is an indication of incipient gear problems.

Gear Excitation.   Gears can be manufactured to such a high degree of precision that very slight imperfections can create abnormal vibration components. These imperfections may either arise during manufacturing or from installation of operation. Mounting inaccuracies may cause otherwise perfect gears to run roughly. Measure-

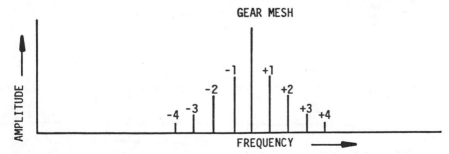

Figure 7-9. Sideband spacing or modulation of a gear set will be symmetrical. Sidebands are equally spaced at the true shaft running speed.

ments of gear error usually reveal a fairly complex pattern of geometric inaccuracies that result in abnormal vibration frequencies.

For vibration analysis of gearbox condition, the lower frequency harmonics are of the greatest interest because these components excite the most destructive drive-train natural frequencies. Higher harmonics, such as tooth-to-tooth gear errors and fluctuations in shaft displacement due to gear tooth flexibility, generate noise rather than vibration in gearboxes.

Backlash. Backlash is an important factor in proper gear operation. All gears must have a certain amount of backlash to allow for the tolerances in concentricity and tooth form. Not enough backlash causes early failure due to overloading. Too much backlash increases the contact force and also reduces the operating life of the gear set.

Abnormal backlash alters the spacing of the sidebands that surround the gear-meshing frequency (Figure 7-10). Rather than main-

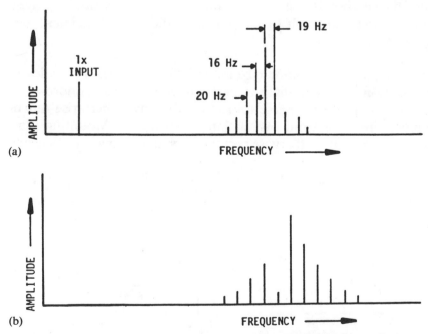

Figure 7-10. Gear defect profiles or signatures: (a) Improper backlash or wear will change the sidegband spacing, (b) Broken teeth will change the amplitude of sidebands.

taining uniform spacing at the shaft running speed, the spacing is erratic.

Monitoring the mechanical condition of gears and gearboxes using vibration-analysis techniques must also consider the unique forces that specific gears generate. For example, a helical gear, by design, generates a high thrust load created by the meshing of the mating gears. Degradation of the helical gear's condition will increase the axial force and its corresponding vibration amplitude.

Narrowband monitoring techniques are ideal for detecting incipient gear problems. A narrowband should be established that includes at least 5 sidebands on either side of the gear-meshing frequency. For example, a gear set with a shaft running speed of 20 Hertz and a meshing frequency at 200 Hertz would have a narrowband with a lower limit set at 100 Hertz (i.e., 20 Hz × 5) and an upper limit at 300 Hertz. This type of narrowband will allow the microprocessor-based data-acquisition unit to automatically report any abnormal increase in the energy generated by the gear set and, therefore, any change in the mechanical condition. This automatic function will not provide the root-cause or failure mode. To determine the actual failure mode will require full, manual signature analysis.

Lubricating oil analysis using spectrographic and wear particle techniques will add useful information about the operating condition of gearboxes and can augment a vibration-monitoring program. However, the added cost normally does not justify the inclusion of these techniques unless a serious problem has been identified by the vibration-monitoring program.

Running Speeds

Every machine-train has at least one true running speed. This is the true rotational speed of the shaft or shafts within the machine. In most cases, there is more than one shaft, and each has its own unique running speed (Figure 7-11).

Since most vibration frequencies are related to the running speed of a shaft within the machine, it is important that every running speed and its unique rotational frequencies be identified. The fundamental vibration frequency or running speed is a primary indica-

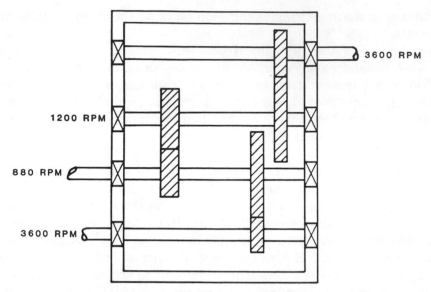

Figure 7-11. Machine-trains can have multiple running speeds. Each shaft or component within a machine will generate a series of frequencies that are unique to that component.

tor of many machine-train problems and should be monitored closely.

A narrowband should be established to automatically monitor and trend each true running speed within the machine-train. It should be noted that the running speeds within a machine do not remain constant. Even constant-speed machines have some variation in the true running speed. This variation is primarily a function of the loading factor. As the load increases on most machines, the running speed decreases. To compensate for this type of speed variation, the narrowband should be established with the lower limit set about 10% below, and the upper limit about 10% above, the calculated or normal running speed. This should be sufficient to compensate for slight variations in the true running speed.

Variable-speed machinery must be handled in a slightly different manner. Many of the microprocessor-based systems provide an alternate method of automatically setting narrowband filters during data acquisition. This method, called "order" analysis, uses a tachometer input to determine the true running speed and then automatically moves the narrowband filters to the correct setting based on the measured running speed. This approach simplifies automatic monitoring of variable-speed machines.

Critical Speeds

All shafts in rotating mechanical systems exhibit potentially damaging radial vibration when operated at certain speeds. These critical speeds are determined by rotational frequencies that coincide with one or more of the shaft-system's natural frequencies. Shafts vibrate at critical speeds even in precisely balanced machines.

Balancing will narrow, but not eliminate, the range of speeds in which vibration builds to a peak. Consequently, a well-balanced machine can operate somewhat closer to the system's critical speeds without being damaged.

Sometimes, the lowest critical speed (called the first or fundamental critical speed) is of the most concern because it is the source of the greatest vibration amplitude. Shafts often are selected for their high stiffness and low mass, which help place the first critical speed above the normal operating-speed range. Rigid, light shafts alone do not, unfortunately, solve all critical-speed problems. For example,

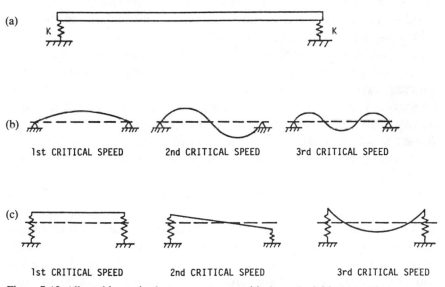

Figure 7-12. All machine-trains have one or more critical speeds. Critical speed is determined by the natural frequency of a specific machine-train and is affected by the rotor and rotor-support stiffness. K is the spring constant.

most fans and blowers are designed to operate just below the first critical speed. As long as the fan or blower is operated at its designed speed, it should not experience a critical-speed problem. In actual operation, fans and blowers are often damaged or destroyed as the direct result of a critical-speed problem. The factor that caused the fan to operate at the first critical speed was plate-out. In normal operation, fans and blowers are subject to a buildup of dirt and other solids contamination on the blades. This increase in mass lowers the first critical speed of the fan. As a result, the fan operating at design speed is now within the first critical speed.

It is not widely realized that critical speed is a rotating-machinery phenomenon and not just a dynamic characteristic of rotating shafts. Any component in a rotating machine that reduces stiffness or increases rotating mass also shifts critical speeds closer to the operating speeds of the machine. This can become a real problem when components are designed or selected without regard to potential process influence on the machine's critical speeds.

Excessive vibration created by operating at critical speeds will immediately decrease as the machine's speed is changed. Therefore, one test method to determine if a critical speed problem exists is to change the operating speed. An increase or decrease in machine speed will drastically reduce both the overall vibration and the true running-speed component.

In addition to the fundamental, or first, critical speed caused by the centrifugal forces of an unbalanced mass, some vibration components have been observed at one-half of the first critical speed. This effect typically is limited to horizontal shafts—indicating that gravity must be one of the causes. There are two primary causes for this secondary critical speed: (1) gravity in combination with imbalance, and (2) gravity in combination with nonuniform bending stiffness of the shaft.

The theory of actual motion and vibration is very complicated, and a comprehensive technical understanding is not necessary for predictive maintenance applications. Critical speeds should be considered on all rotating machines that are included in the program. Special attention to criticals should be paid on any machine-train that has overhung or cantilevered components (e.g., fans) or machines with high rotating mass and low foundation mass. Paper machines are classic examples of the high rotating mass and low support-structure mass. The majority of the machine's mass is in the rotating components (i.e., the rolls). This mass is supported by minimum structure and is, therefore, highly susceptible to critical-speed problems.

Mode Shape

All rotating shafts have multiple mode shapes created by either the critical speed or process-induced forcing functions (Figure 7-13). Since most process machinery use relatively flexible shafts, the shafts tend to deflect and operate in a shape or mode rather than rotate on its true centerline. The first mode of operation is a radial offset from the shaft's true centerline. In this mode, the shaft actually rotates cylindrically around the normal, or static, centerline and does not have a true null, or zero, point between the bearing supports. In this mode, the shaft remains relatively straight but rotates offset or displaced around the true centerline. Single plane imbalance and other forcing functions create this mode of operation. The

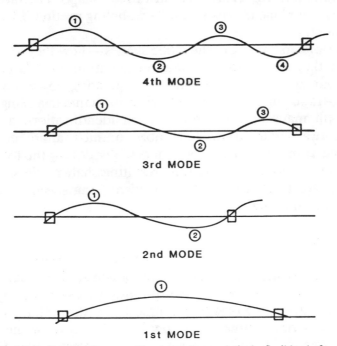

Figure 7-13. Mode shapes. Most machine-trains have relatively flexible shafts. Abnormal operation will force the shaft to flex and assume a specific mode shape.

true running speed vibration component will be excited by this mode of operation.

The second mode shape taken by a rotating shaft is conical rather than cylindrical. In this mode, the shaft is deformed into a mode that resembles an "S" shape with one null, or zero, point between the bearing supports. Therefore, for every revolution of the shaft, two high spots can be observed in the vibration signature. This mode, normally associated with misalignment, creates a vibration component at two times the running speed, or 2X. Other forcing functions within the machine-train or process system can also create this mode shape. For example, out-of-phase imbalance, bent shafts, aerodynamic instability, and many other failure modes can create the second mode shape.

The third mode shape of a rotating shaft is a compound deformation of the shaft. In this mode, the shaft has two null, or zero, points between the bearing supports. This mode creates a vibration frequency component at three times (3X) the running speed. Many forcing functions can create the third mode shape. The most common are multiplane imbalance and a wobbling motion of the rotating element.

A clear understanding of the mode shapes—how they are created and how they appear in the vibration signature—will greatly improve your ability to understand the operating condition of the machine-trains included in the program. Understanding mode shapes will make analysis and problem identification easier and much more accurate. Many of the more common failure modes will excite one or more of these mode shapes. Visualizing the forces that these failure modes create and the resulting shaft mode shape will provide a great deal of assistance in identifying specific problems within a machine-train.

Resonance

All machines have a natural frequency that is the combination of all machine and structural frequency components. If this natural frequency is excited by one or more running speeds or by a defect in the machine-train, the machine's support structure will magnify the energy and serious damage may occur. Abnormal resonance of a machine's natural frequency is one of the most destructive vibration forces that you will ever encounter. It is capable of creating catastrophic failure of machine housings and support structures.

Normally, the frequencies associated with resonance are low; in some cases, they are below the monitoring capabilities of your predictive maintenance instrumentation. To monitor machine resonance frequencies, your instrumentation must be able to separate frequencies in the 1 to 10 Hertz range from the DC noise that usually limits monitoring these bands.

Sometimes this resonance can be transmitted to adjacent machines or plant equipment. It can also mask the vibration signature of other critical machine components. If resonance is suspected, it should be verified and eliminated as quickly as possible.

Preloads and Induced Loads

Designed and dynamically induced loading on rotating shafts is one of the most common yet misunderstood of all machine behaviors. They also are a major contributor to machine failure.

Preload is defined as a directional force that is applied to a rotating shaft by design. An example of preload is the side-load created by a belt drive assembly. Most machines have at least one designed preload that creates a directional force that is not compensated for by an equal and opposite force. Gravity is one form of preload. All machines have this unbalanced force to overcome during normal operation.

Induced load is also an unbalanced directional force within a machine. However, in this case, the force is created by the dynamic operation of the machine or system. An example of induced load is the aerodynamic instability created by restricting air flow through a fan or blower. All bladed or vaned machines (e.g., pumps, compressors, fans) are susceptible to this type of abnormal loading.

The result of both preloads and induced loads is the deflection of the rotating shaft into one quadrant of the bearings or into one of its mode shapes. This results in a nonlinear resistance, in that the spring constant of the bearings is much higher opposing the force than it is perpendicular to the load. This will cause premature bearing failure and can cause other serious damage to the machine-train.

Understanding these loads and how they affect the machine-train is important for two reasons. First, it will enable you to locate the primary data point in the plane opposing the potential induced load. This will ensure the early detection of an incipient problem. Second, it will provide assistance in diagnosing developing machine problems.

Preloads and induced loads do not necessarily cause machine malfunctions. In some cases, they tend to stabilize the shaft, bearings, and rotating elements. However, if excessive loads are applied to machinery, serious problems can, and in most cases will, develop very quickly. In extreme cases, bent shafts, cracked rotating elements, cracked couplings, and other catastrophic machine problems have developed.

Process Variables

Most mechanical equipment is designed to perform a function within a process system. Therefore, a predictive maintenance program cannot rely strictly on monitoring the vibration data from these mechanical systems. Variations in the process envelope have a direct effect on the operating condition of most mechanical equipment. Pumps, compressors, fans, and other mechanical equipment rely on minimum suction pressures to operate and are limited to the maximum discharge pressure (TDH) that they can generate (Figure 7-14). Variations in suction pressure (NPSH) and the discharge-

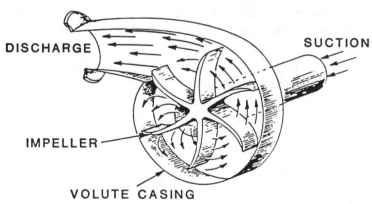

Figure 7-14. Induced process loads. Changes in the process will directly change the mechanical condition of a machine. What happens to the mechanical forces applied to a pump or fan if the suction or discharge condition is changed?

pressure demands of process systems can prevent the mechanical equipment from operating within an acceptable environment and can cause catastrophic failure of the mechanical system. Many of the problems that cause premature failure of mechanical systems are the direct result of process-induced loads. A large percentage of machinery imbalance problems are caused by hydraulic or aerodynamic instability created by process restrictions (Figure 7-15). In these instances, the system is demanding that the mechanical equipment operate outside its capabilities. A pump, for example, cannot deliver product if the discharge valve is closed. Nor can the pump continue to operate in this condition.

Even though vibration monitoring and analysis techniques detect the abnormal vibration caused by process-induced instability, the addition of recorded process data greatly enhances a predictive maintenance program. All of the process parameters that directly affect the operating of mechanical equipment should be acquired and recorded as part of the routine data-acquisition process. In addition to supporting the vibration-analysis function, long-term

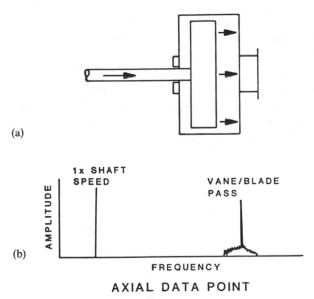

Figure 7-15. Effects of process loads. (a) An increase in suction-lift conditons will attempt to pull the rotor out of the fan housing, (b) The vibration signature will indicate this type of change by an increase in the running speed (ix) and vane-pass frequency components.

trends of these process variables will often identify a potentially serious system problem.

Most of the microprocessor-based systems support at least one method of recording process variables as part of the routine data-acquisition process. Some will allow direct acquisition of process variables from installed plant instrumentation. Others will allow manual entry of process data.

Blades and Vanes

Machinery that use blades or vanes have an additional frequency that should be routinely monitored. The frequency, called blade- or vane-pass, represents the frequency created by the blades or vanes passing a reference point, usually the vibration transducer. The passing frequency can be calculated by multiplying the number of blades or vanes by the true running speed of the machine's shaft.

The amplitude and profile of the passing frequency vary with the load. Therefore, it is important to record the actual operating load as part of the data-acquisition process. In a normal machine, the passing frequency should be a low-level, distinct peak at the calculated frequency.

If process-induced instability is present, the passing frequency will increase in amplitude, and modulations or sidebands around the passing frequency will develop.

A narrowband should be established to automatically monitor the blade- or vane-passing frequency. The lower limit should be set about 10% below, and the upper limit about 10% above, the calculated passing frequency to compensate for variations in speed and to capture the sidebands that will be created by instability.

Belt Drives

Machine-trains that use belt drive assemblies have an additional set of frequencies that should be monitored. All belt drives have a belt-pass frequency that identifies the operating condition of the drive system. This unique frequency is generated by the true running speed of the belts and can be calculated for any drive assembly (Figure 7-16).

$$\text{Belt Length} = 2L + (d_1 \times 3.1416)/2 + (d_2 \times 3.1416)/2$$

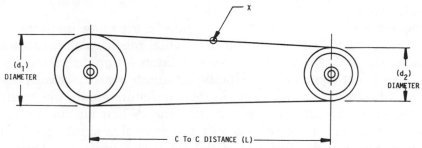

Figure 7-16. Belt drives create a unique set of frequencies that can be used to identify defects.

Calculate the belt-pass frequency by:

Belt-pass frequency = $(d_1 \times 3.1416 \times N) /$ Belt Length

Where: d_1 = Drive sheave diameter, in inches
d_2 = Driven sheave diameter, in inches
L = Center-to-center distance, in inches
N = True rotating speed of drive shaft, in Hertz.

Belt-pass frequency is a good means for identifying misalignment, excessive induced load, and other failure modes that are associated with the drive assembly.

A narrowband should be established to automatically monitor the belt-pass frequency. The lower limit should be set at 10% below the calculated pass frequency. The upper limit should be set at a multiple of the calculated frequency that is equal to the number of belts; for example, a drive with 10 belts would set the upper limit at 10 times the calculated pass frequency.

Common Failure Modes

Many of the common defects, or failure modes, in mechanical equipment can be identified by understanding their relationship to the true running speed of a shaft within the machine-train. The following section will discuss the general definitions of the most common machine-train failure modes (i.e., imbalance, misalignment, bent shaft, mechanical looseness or rub). These definitions are guidelines and should not be accepted as being true in all cases.

Imbalance

Imbalance is probably the most common failure mode in mechanical equipment. The assumption that actual mechanical imbalance must exist to create an imbalanced condition within the machine is incorrect. Aerodynamic or hydraulic instability also can create massive imbalance in the machine. In fact, all failure modes will create some form of imbalance in the machine. When all failures are considered, the number of machine problems that are the result of actual mechanical imbalance of the rotating element is relatively small.

Imbalance can take many forms in the vibration signature. In almost every case, the fundamental or running speed component will be excited; this is the dominant amplitude. However, this condition can also excite multiple harmonics or multiples of running

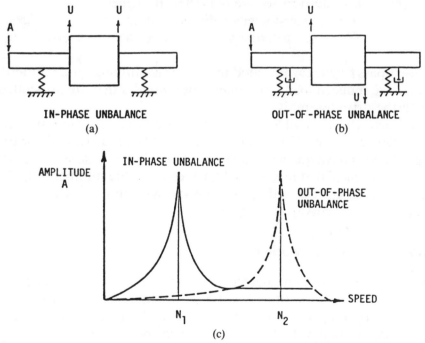

Figure 7-17. Single plane imbalance. (a) In-phase imbalance will create an increase in the running speed (1X) frequency component, (b) Out-of-phase imbalance will create an increase in the second order (2X) component, (c) Example of vibration signature created by both in-phase and out-of-phase imbalance. A = Amplitude. U = Unbalance. N = Running speed. $N_2 = 2X$ Running speed.

speed. The number of harmonics and their amplitude have a direct correlation to the number of planes of mechanical imbalance and their phase relationship.

For a single-element rotating machine, a narrowband should be established to monitor the fundamental or true running speed frequency component. For multiple-element rotating machines, the narrowband should monitor the true running speed, and the number of harmonics should equal the number of rotating elements (Figure 7-18). As in earlier examples, the narrowband limits should allow for slight variations in running speed.

Misalignment

This condition is virtually always present in machine-trains. Generally we assume that misalignment exists between two shafts connected by a coupling, v-belts, or other intermediate drives. Misalignment can also exist between the bearings of a solid shaft or at other points within the machine (Figure 7-19).

The presentation of misalignment in the vibration signature depends on the type of misalignment. There are two major classifications of misalignment: parallel and angular.

Parallel misalignment is present when two shafts are parallel to

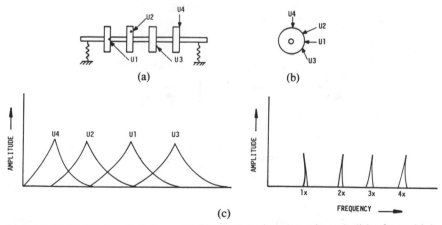

Figure 7-18. Multi-plane imbalance. (a) Single-plane imbalance in each disk of a multiple disk rotor will create an increase in a harmonic of the shaft running speed, (b) four unique points of imbalance will create four unique harmonics at 1X, 2X, 3X and 4X of the shaft running speed. U_1, U_2, U_3, U_4 are points of unbalance.

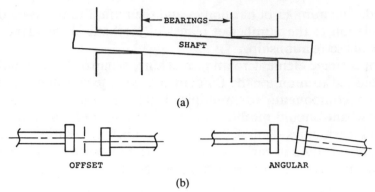

Figure 7-19. Types of shaft misalignment: (a) at bearings, (b) at a coupling.

each other but have axes which do not coincide (Figure 7-19 and 7-20). This type of misalignment generates a radial vibration and duplicates the second mode shape. In other words, it generates a radial vibration at two times (2X) the true running speed of the shafts.

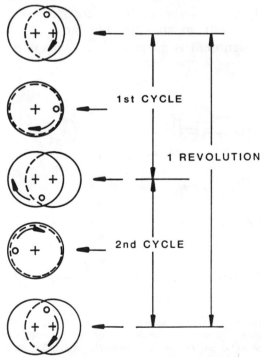

Figure 7-20. Angular misalignment at a coupling will result in high axial vibration at 1X of shaft running speed.

Angular misalignment exists when the shafts are not parallel to each other (Figures 7-19 and 7-21). This type of misalignment generates axial (i.e., parallel to the shaft) vibration. Since this form of misalignment can duplicate any of the mode shapes, the resultant vibration frequency can be at the true running speed of the shaft, two times (2X) the running speed, or three times (3X) the running speed. The key indicator is an increase in axial vibration.

Bent Shaft
A bent shaft creates an imbalance or misaligned condition within the machine-train. It should be treated in the same manner as these common failure modes.

Mechanical Looseness
Looseness can create a variety of patterns in the vibration signature. It usually creates primary frequency components at 50% of running speed and can generate multiple harmonics of this primary compo-

SHAFT ANGULAR MISALIGNMENT

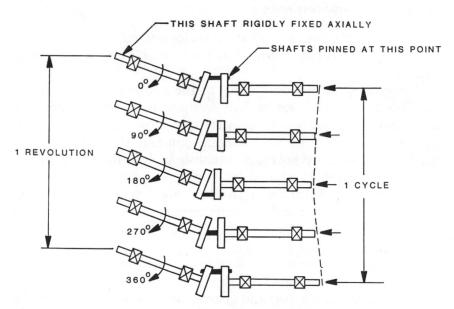

SHAFT OFFSET MISALIGNMENT

Figure 7-21. Parallel misalignment at a coupling will result in high radial vibration at 2X of shaft running speed.

nent. In other words, there will be a frequency component at 50%, 150%, 250%, and so forth. Sometimes, the fundamental (or true) running speed component (1X) is excited. In almost all cases, there are multiple harmonics with almost identical amplitudes.

Mechanical Rub

Many machine-trains are susceptible to mechanical rub. This failure mode may be a shaft grinding against the babbitt of a sleeve bearing, the rollers in a rolling-element bearing against the races, or some part of the rotor against the machine housing. In each case, the vibration signature displays a low-amplitude peak, normally between 1.0 and 10.0 Hertz. This extreme low-frequency peak has high amplitude and is accompanied by a set of lesser peaks at about 25% and 40% of the shaft's running speed. Machine failure, when the defect is present, is highly probable. Note that not all vibration-monitoring system can detect this defect. Many have a low-frequency cutoff at 10.0 Hertz and cannot capture any frequencies below this level.

Monitoring Requirements

In addition to the common characteristics and failure modes, each machine type has unique requirements for monitoring its operating condition.

After gaining an understanding of the common characteristics and failure modes of mechanical equipment, the next step is to determine where and how to monitor specific plant machinery. For each machine-train, use a consistent, common-shaft approach to locate measurement points and establish analysis parameter sets.

Measurement points should be numbered sequentially, starting with the outboard driver bearing and ending with the final outboard bearing of the driven machine component. In addition, a consistent numbering sequence that identifies the orientation (e.g., vertical, horizontal, axial) should be used.

The term "common shaft" refers to each continuous shaft in the machine-train. For example, in an electric-motor-driven, single-reduction gearbox, the motor and gearbox input shaft is considered a common shaft. Even though the shaft is coupled, all forces acting on the extended shaft, and any resultant vibration, are transmitted throughout the shaft.

This approach to setting up a data base, monitoring operating conditions, and analyzing incipient problems ensures immediate identification of the location of a particular data point during acquisition and analysis of a machine-train. This approach also enables the analyst to evaluate all parameters that affect each component of the machine.

Understanding the specific location and orientation of each measurement point is critical when diagnosing incipient machinery problems. The vibration signature is a graphical representation of the actual dynamic forces within the machine. Without knowing the location and orientation, it will be difficult (if not impossible) to correctly identify the incipient problem. The orientation of each measurement point should be considered carefully also during the data-base setup. There is an optimum orientation for each measurement point on every machine-train in your program. In almost all cases, each bearing cap will require two radial measurements (i.e., perpendicular to the shaft) to properly monitor the machine's operating condition. There should also be at least one axial measurement (i.e., parallel to the shaft) on every common shaft.

These measurement points should be oriented to monitor the worst possible dynamic force and vibration. As an example, a belt-driven fan has its dominant force created by the belt tension; therefore, the worst vibration usually is in the radial direction of the belt drive. To monitor the worst vibration and gain the earliest possible detection of an incipient problem, the primary radial measurement point should be between the shaft and a point opposite from the belt drive (Figure 7-22). A secondary radial measurement should be

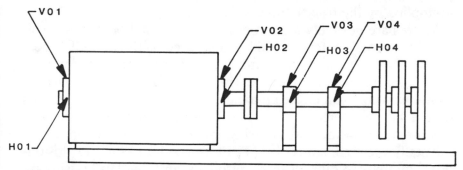

Figure 7-22. Typical measurement point locations. Most machine-trains will require 2 radial measurement points at each bearing. An axial measurement is required at bearings subjected to axial loads. V and H represent vertical and horizontal measurement points, respectively.

taken at 90 degrees to the primary. This secondary measurement provides a comparative energy reading and helps the analyst determine the actual force vector within the machine.

In the above example, axial readings on both the motor and fan shafts are very important as well. A common failure mode of belt-driven units is misalignment. If the sheaves are not in the same plane, the belt tension attempts to self-align, creating axial movement in the two shafts. The axial measurements can detect these abnormal forces and identify the alignment problem.

Electric Motors

Electric motors are often used as the prime mover for process and manufacturing machinery. Depending on size and manufacturer, they may use either sleeve or rolling-element bearings. They rarely have thrust bearings, so they are susceptible to abnormal axial movement when coupled to process equipment that can create thrust load.

In addition to the common failure modes, electric motors are prone to several unique problems. These include: loss of insulation, loose or broken rotor bars, loose poles, and electrical shorts. These failure modes can be monitored by including narrowbands that monitor line frequency (e.g., 60 Hertz) and its harmonics (e.g., 120, 180). If electrical problems exist, the line frequency and harmonics would clearly indicate a problem. Loose rotors or broken rotor bars can be detected by monitoring the current to the motor. If the condition exists, the motor's slip frequency is clearly displayed as sidebands on either side of the line frequency. Loose poles show as a pole-pass frequency or at a frequency equal to the number of poles multiplied by the true running speed.

Motors are normally used in either a horizontal or vertical position. Horizontal motors should be monitored using two radial measurements points on the inboard and outboard bearings. An axial reading is not necessary unless the motor is driving a machine component that can create an axial or thrust load. Vertical motors should be monitored in the same manner but require an axial measurement on the lower bearing in the upward direction.

Totally enclosed, fan-cooled (TEFC) motors and some explosion-proof motors are difficult to monitor. Because the fan housing on these motors encloses the outboard bearing cap, the best method of

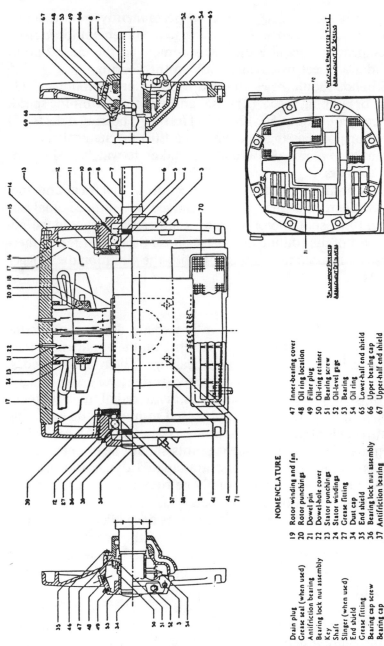

NOMENCLATURE

3 Drain plug	19 Rotor winding and fan
4 Grease seal (when used)	20 Rotor punchings
5 Antifriction bearing	21 Dowel pin
6 Bearing lock nut assembly	22 Dowel-hole cover
7 Key	23 Stator punchings
8 Shaft	24 Stator windings
9 Slinger (when used)	27 Grease fitting
10 End shield	34 Dust cap
11 Grease fitting	35 End shield
12 Bearing cap screw	36 Bearing lock nut assembly
13 Bearing cap	37 Antifriction bearing
14 End-shield bolts	38 Grease seal (when used)
15 Frame	39 Bearing cap
16 Air-deflector screw	41 Conduit box
17 Air deflector	42 Conduit-box screw
18 Rotor key	46 Inner-bearing-cover screw

47 Inner-bearing cover	
48 Oil ring location	
49 Filler plug	
50 Oil-ring retainer	
51 Bearing screw	
52 Oil-level gage	
53 Bearing	
54 Oil ring	
65 Lower-half end shield	
66 Upper bearing cap	
67 Upper-half end shield	
68 Dowel pin	
69 Bearing-cap bolt	
70 Screen (weather-protected, Type I)	
71 Louver (splashproof)	

Figure 7-23. Cross-section of a typical electric motor. Each component within the motor can create unique vibration frequencies.

acquiring outboard bearing data is to permanently mount transducers on the bearing cap and wire them to a convenient location. If this is not possible, acquire the measurement at the closest point that provides a direct mechanical link to the bearing.

Narrowbands should be established to monitor the following: (1) imbalance or running speed, (2) misalignment or 2X running speed, (3) electrical problems at 60, 120, 180 Hertz, (4) bearing defects, (5) loose poles or pole-passing frequency, and (6) mechanical rub.

A current loop reading should also be taken to watch for loose or broken rotor bars.

Infrared scanning provides early detection of electrical and mechanical problems that may not be detected by vibration analysis. At a minimum, spot motor-winding temperatures should be acquired as part of the routine data-acquisition process. Infrared thermometers can be used in conjunction with the vibration data logger for these data.

Gearboxes

Gearboxes are widely used as an intermediate drive to either increase or decrease driver speed. Depending on the application, they may use a variety of gear types and bearings. Their failure modes, other than the common modes discussed earlier, vary accordingly.

Regardless of the gearbox internals, two radial measurement points on each bearing cap should be used to monitor its operating condition. If helical gears are used, an axial reading on each shaft also is required. The measurement points should be oriented in the direction opposite from the worst anticipated dynamic force and resulting vibration, which will usually be opposing the preload and thrust loads generated by the gear set.

Multiple reduction, or increase gearboxes have idler shafts that, in many cases, are not accessible from outside the gearbox. In these situations, axial readings at the point closest to the idler shafts should provide reliable data if the selected point provides a direct mechanical link to the shaft or bearing housing.

Narrowbands should be established for each shaft to monitor the following: (1) imbalance, (2) misalignment, (3) gear-meshing frequency, (4) bearing defects, and (5) mechanical rub.

Bearing-cap temperature and motor amp load should be acquired as part of the measurement set. Periodic lubricating oil analysis, using spectrographic and wear particle techniques, would improve the program's ability to detect incipient problems.

Fans and Blowers

The variety of designs of fans and blowers is almost infinite. They generally fall into two classes: centerline and cantilevered (i.e., over-hung). Both classifications are generally designed to operate just below their first critical speed, and therefore, are prone to severe imbalance created by the critical speed.

The cantilevered design also is susceptible to aerodynamic insta-bility and induced loads. This is primarily the result of the high mass of the rotating element and the overhung configuration of the bear-ing support structure. All fans and blowers should be monitored for the common failure modes and for process-induced instability. Blade-pass frequency is a primary indicator of condition.

Belt-driven units are prone to misalignment, so they should be

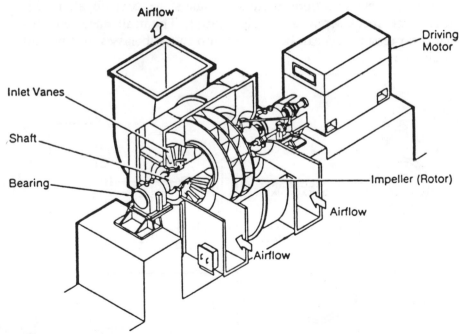

Figure 7-24. Cross-section of a typical centerline fan. Monitor each component's unique vibration signature.

watched closely. Two radial measurement points on each bearing cap oriented to oppose the worst dynamic force and at least one axial measurement are required to monitor fan operating condition (Figure 7-25).

Narrowbands should be established to monitor the following: (1) imbalance, (2) misalignment, (3) bearing defects, (4) blade-pass frequency, (5) aerodynamic instability, and (6) mechanical rub.

Attention should be paid to the suction and discharge locations. Restrictions in either of these will try to force the rotating element in the opposite direction. Measurement points opposing suction and discharge direction will be the correct monitoring direction.

Process data such as suction pressure, discharge pressure, and motor amp load should be included in the measurement set. Bearing cap temperatures would also assist in early detection of bearing problems.

Compressors

Like fan designs, the variety of compressor designs is almost infinite. The major classifications include: single-stage centrifugal, multistage centrifugal, screw, and reciprocating. The multistage centrifugal compressor class can be divided into two subclasses: in-line and bullgear.

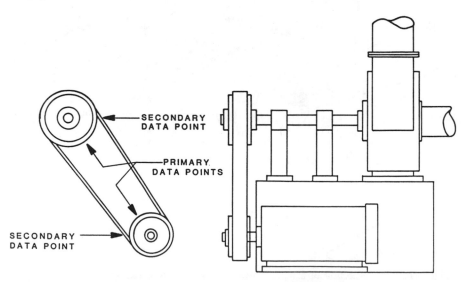

Figure 7-25. Typical measurement points for fans.

Both rolling-element, sleeve and tilting-pad bearings are used in industrial compressors. Monitoring parameters should be established according to bearing type.

Single-stage centrifugal compressors are similar to blowers and can be monitored in the same manner. Vane-pass should be a primary indicator of condition. Narrowbands should be established to monitor: (1) imbalance, (2) misalignment, (3) vane-pass frequency, (4) bearing defects, (5) aerodynamic instability, and (6) mechanical rub.

Multistage in-line compressors are similar to multistage centrifugal pumps and can be monitored in the same manner. Narrowbands should monitor the same failure modes as listed above for single-stage centrifugal compressors, as well as each vane-pass frequency.

Multistage bullgear compressors should be treated as a combination of a gearbox and a pump. By design, these compressors have a large, helical gear, a bullgear that drives several smaller gears mounted on impeller pinion shafts (Figure 7-27). The pinion shaft speeds for this type of compressor are typically in the 30,000 to 50,000 rpm range and should be monitored closely. The pinion shafts normally have tilting-pad bearings that will generate a passing frequency if abnormal clearance or alignment exists. Most bullgear compressors have displacement transducers permanently mounted to monitor the pinion shafts. Data should be acquired from these sensors and from casing sensors.

Two radial measurement points on each bearing cap of the bullgear and each pinion shaft should be used to monitor the compressor's operating condition. Because the compressor uses helical gears, axial measurements of each bearing cap are also required. For each shaft, narrowbands should be established to monitor the following: (1) imbalance, (2) misalignment, (3) gear mesh, (4) vane-pass frequency, (5) bearing defects, (6) aerodynamic instability, and (7) mechanical rub.

Screw compressors may use either rolling-element or sleeve bearings and are susceptible to aerodynamic instability. They normally have extremely close axial tolerances that allow no more that 0.5 mils of axial movement before the rotors contact. Axial measurements on both rotors or screws are absolutely critical for these machines. Two radial and one axial measurement on each bearing cap should be used on screw compressors. The rotor-meshing and

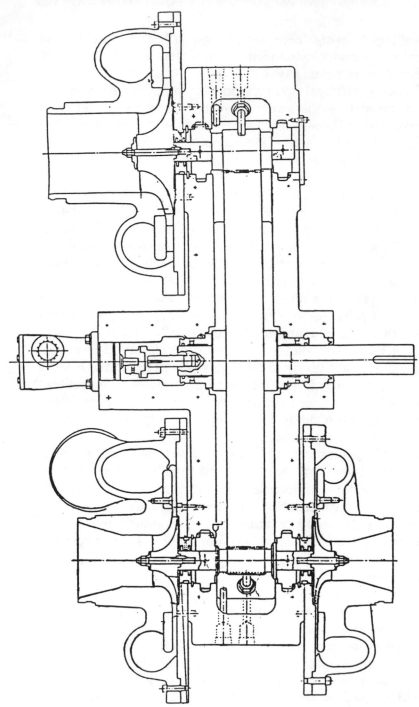

Figure 7-26. Cross-section of bullgear compressor. This type of machine is subject to process induced instability. Monitor the vane-pass, pad-pass gear mesh frequencies closely.

axial readings are the primary indicators of abnormal operation. For each rotor, narrowbands should be established to monitor the following: (1) imbalance, (2) misalignment, (3) rotor-meshing, (4) bearing defects, (5) aerodynamic instability, and (6) mechanical rub.

Reciprocating compressors generate forces and resultant vibration at frequencies different from rotating machines. The second harmonic (2X), rather than the true crank-shaft speed, is usually dominant. In addition to the two radial measurement points on each bearing cap, measurements should be taken on the cylinder wall (to detect rubbing) and near the suction and discharge valves (to detect valve problems). A full set of process parameters should be acquired for all compressors. Narrowbands should monitor the following: (1) imbalance of the crank shaft, (2) misalignment of the crank shaft, (3) mechanical rub in cylinder walls, (4) valve defects, (5) bearing defects, and (6) looseness.

Regardless of design, compressors are prone to process-induced problems. All the specified data are necessary to determine their actual operating condition.

Process parameters provide valuable information about the operating condition of all compressors. All envelope parameters (e.g., pressures, temperatures, amp load on motor, bearing-cap temperatures) should be included in the data set. Interstage data on multistage compressors, especially reciprocating units, are critical to analysis.

Periodic lubricating oil analysis, using spectrographic or wear particle techniques, can provide early warning of incipient problems.

Generators

Generators are usually supplied with fluid-film or sleeve bearings and should be monitored with two radial measurement points on each bearing cap. Because they are prone to end-play or axial movement of the entire rotor assembly, at least one axial measurement is required. Narrowbands should monitor the following: (1) imbalance, (2) misalignment, (3) bearing defects, (4) rotor instability (normally 3X running speed), (5) electrical defects, and (6) mechanical rub.

Infrared scanning of the generator can provide early warning of incipient problems that may not be detected by vibration analysis.

Pumps

The variety of pumps is almost infinite. They should be monitored in the same manner as fans and blowers. The vane-pass frequency is the primary indicator of process problems and should be monitored closely. Radial and axial measurement points should be oriented to monitor process-induced loads. The worst radial force should be opposing the discharge on end-suction pumps, and be in line with the suction and discharge on split-case pumps. End-suction pumps should also be monitored using axial measurement points to detect suction problems.

Multiple-stage centrifugal pumps can, depending on the design, create high thrust or axial loads. Multistage in-line pumps, with all impellers facing in the same direction, must be monitored closely for any increase in axial movement or load. Opposed impeller designs normally balance the axial load and need not be monitored for axial loads.

Narrowbands should monitor: (1) imbalance, (2) misalignment, (3) each vane-pass frequency, (4) bearing defects, (5) hydraulic instability, and (6) mechanical rub.

A full set of process parameters is required on pumps because they, like compressors, are highly susceptible to process-induced problems. These measurements should include pressures, temperatures, flow, motor amp load, and bearing temperatures.

Continuous Process

Most manufacturing and process plants use a variety of complex, continuous-process mechanical systems that should be included in the predictive maintenance program. Included in this classification are: paper machines, rolling mills, can lines, printing presses, dyeing lines, and many more. These systems can be set up, monitored, and analyzed in the same manner as simpler machine-trains such as pumps and fans. The initial data-base setup for the complex machinery would require more effort, but the same principles apply.

Each system should be evaluated to determine the common shafts that make up the total machine-train. Using the common-shaft data, evaluate each shaft to determine the unique mechanical motions and dynamic forces that each one generates, the direction of each force, and the anticipated failure modes. This information can then be used to determine measurement point locations and the narrowbands that are required to routinely monitor the machine's operating condition.

Narrowband selection is dependent on the operating dynamics of each machine. The same methods used for simpler machinery should be employed. When establishing the narrowbands, each shaft should be treated as the basic unit.

NONMECHANICAL EQUIPMENT

A certain portion of critical plant equipment is nonmechanical. This equipment can be divided into two classifications: electrical and process.

Each of these classifications have common failure modes that should be monitored on a regular basis to maintain plant performance. A vibration-based program normally does not address the monitoring requirements of electrical equipment. Therefore, an alternate or secondary means of regular monitoring is required for the predictive maintenance program.

Electrical Equipment

The temperature profile of electrical equipment is the primary means of monitoring their operating condition. Thermal imaging, using thermographic techniques, is the recommended technique required to regularly monitor their condition. However, as discussed in previous chapters, the cost of implementing a full-capability thermographic program is, in most cases, prohibitive. The most cost-effective method of handling this requirement is the use of an outside contractor who can provide the monitoring capabilities required.

Exceptions to the use of thermographic techniques for electrical equipment are those systems that rely on dielectric oils to function. In this instance, tribology provides a better understanding of the equipment's condition. As with thermal imaging, this information can be more cost-effectively acquired using outside sources.

Process Equipment

Process equipment, while considered static (i.e., no moving parts), are true dynamic systems. For example, heat exchangers, filters, and other process equipment perform work. Therefore, these systems should be monitored using process efficiency calculation techniques to determine their operating condition.

Heat exchanger performance can be limited by several failure modes, including fouling, leaks, and flow reductions. Each of these will limit the heat-transfer efficiency of the unit. By regular monitoring of the process variables of flow, pressure, and temperature, each of the failure modes can be detected and quantified.

Filtration systems can be affected by these failure modes. Blockage is most often created by fouled filtration media. With regular monitoring of the differential pressure across the media and other process variables, incipient problems can be detected and isolated to the root-cause.

There are numerous nonmechanical systems in today's plants. Each has its own unique failure modes. By understanding these failure modes, you can establish the correct monitoring techniques that should be used as part of your program.

Chapter 8

Data-Base Development

Once the most cost-effective methods to monitor your plant machinery are selected, the next step in establishing a predictive maintenance program is the creation of a comprehensive data base.

ESTABLISHING DATA-ACQUISITION FREQUENCY

During the implementation stage of a predictive maintenance program, all classes of machinery should be monitored to establish a valid baseline data set. Full vibration signatures should be acquired to verify the accuracy of the data-base setup and to determine the initial operating condition of the machinery.

Since a comprehensive program includes trending and projected time-to-failure, multiple readings are required on all machinery to provide sufficient data for the microprocessor to develop trend statistics. During this phase, measurements normally are taken every two weeks.

After the initial analysis of the machinery, the frequency of data collection will vary depending on the classification of the machine-trains. Class I machines should be monitored on a two- to three-week cycle, Class II on a three- to four-week cycle, Class III on a four- to six-week cycle, and Class IV on a six- to ten-week cycle.

This frequency can, and should, be adjusted for the actual condition of specific machine-trains. If the rate of change of a specific machine indicates rapid degradation, you should either repair it or (at least) increase the monitoring frequency to prevent catastrophic failure.

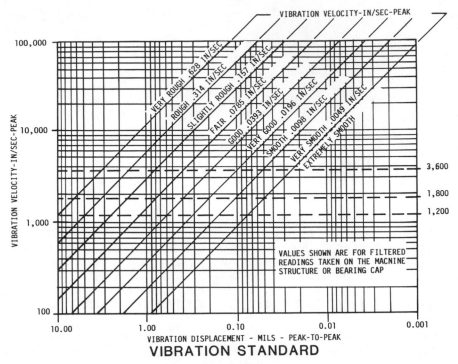

Figure 8-1. Casing vibration standards.

The recommended data-acquisition frequencies will ensure prevention of most catastrophic failures. Less frequent monitoring would limit the ability of the program to detect and prevent unscheduled machine outages.

To augment the vibration-based program, you should also schedule the non-vibration tasks. Bearing-cap, point-of-use infrared measurements, visual inspections, and process parameters monitoring should be conducted in conjunction with the vibration data acquisition.

Full infrared imaging, or scanning, of the equipment included in the vibration-monitoring program should be conducted on a quarterly basis. In addition, full thermal scanning of critical electrical equipment (e.g., switch-gear, circuit breakers) and all heat transfer systems, (e.g., heat exchangers, condensers, process piping) that are not in the vibration program should be conducted quarterly.

Lubricating oil samples from all equipment included in the program should be taken on a monthly basis. At a minimum, a full spectrographic analysis should be conducted on these samples. Wear particle and other analysis techniques should be made on an "as-needed" basis.

SETTING UP ANALYSIS PARAMETERS

Using the information developed in Chapter 7, the next step in establishing the program's data base is to set up the analysis parameters used to routinely monitor plant equipment. Each of these parameters should be based on the specific machine-train requirements discussed above.

The analysis parameter set for nonmechanical equipment usually consists of the calculated values derived from measuring the thermal profile or process parameters. Each equipment or system classification has its own unique analysis parameter set.

BOUNDARIES FOR SIGNATURE ANALYSIS

All vibration-monitoring systems have finite limits on their resolution, which is the ability to graphically display the unique frequency components that make up a machine's vibration signature. The upper limit (F_{max}) for signature analysis should be set high enough to capture and display enough data so the analyst can determine the operating condition of the machine-train, but no higher. Since most microprocessor-based systems are limited to 400 lines of resolution, selection of excessively high frequencies can severely limit the diagnostic capabilities of the program.

To determine the impact of resolution, calculate the display capabilities of your system. For example, a vibration signature with a maximum frequency (F_{max}) of 1000 Hertz taken with an instrument capable of 400 lines of resolution would result in a display in which each displayed line will be equal to 2.5 Hertz, or 150 RPM. Any frequencies that fall between 2.5 and 5.0 (i.e., the next displayed line) would be lost.

DEFINING ALERT AND ALARM LIMITS

The method of establishing and using alert/alarm limits varies depending on the particular vibration-monitoring system that you select. These systems use either static or dynamic limits to monitor, trend, and alarm measured vibration. In this book, no attempt is made to define the different dynamic methods of monitoring vibration severity. It does, however, provide a guideline for the maximum limits that should be considered acceptable for most plant mechanical equipment.

The systems that use dynamic alert/alarm limits are based on the assumption that the rate of change of vibration amplitude is more important than the actual level. Any change in the vibration amplitude is a direct indication that there is a corresponding change in the machine's mechanical condition. However, there should be a maximum acceptable limit—an absolute fault.

The accepted severity limit for casing vibration is 0.628 ips-Peak (velocity). This is an unfiltered broadband value and normally represents a bandwidth between 10 and 10,000 Hertz. This value can be used to establish the absolute fault or maximum vibration amplitude for broadband measurement on most plant machinery. The exception would be machines with running speeds below 1200 RPM or above 3600 RPM.

Narrowband limits, the discrete bandwidths within the broadband, can be established using the following guideline. Normally 60 to 70% of the total vibration energy occurs at the true running speed of the machine. Therefore, the absolute fault limit for a narrowband established to monitor the true running speed would be 0.42 ips-Peak. This value can also be used for any narrowbands established to monitor frequencies below the true running speed.

Absolute fault limits for narrowbands established to monitor frequencies above running speed can be calculated using the 0.42 ips-Peak limit established for the true running speed. For example, the absolute fault limit for a narrowband created to monitor the blade-passing frequency of fan with 10 blades would be set at 0.42/10, or 0.042.

Narrowbands designed to monitor high-speed components (i.e., above 1000 Hertz) should have an absolute fault of 3.0 g-Peak (acceleration).

Rolling-element bearings have, based on factor recommendations, an absolute fault limit of 0.01 ips-Peak. Sleeve or fluid-film bearings should be watched closely. If the fractional components that identify oil whip or whirl are present at any level, the bearing is subject to damage, and the problem should be corrected.

Nonmechanical equipment and systems normally have an absolute fault limit that specifies the maximum recommended level for continued operation. Equipment or systems vendors are, in most cases, able to provide this information.

TRANSDUCERS

The type of transducers and data-acquisition techniques that you select for your program is the final critical factor that can determine the success or failure of the program. Their accuracy, proper application, and mounting will determine whether or not valid data are collected.

The optimum predictive maintenance program developed in earlier chapters is predicated on vibration analysis as the principal technique for the program. This is most sensitive to problems created by the use of the wrong transducer or mounting technique.

There are three basic types of vibration transducers that can be used for monitoring the mechanical condition of plant machinery: displacement probes, velocity transducers, and accelerometers. Each has specific applications within the plant. Each has limitations.

Displacement Probes

Displacement, or eddy-current, probes are designed to measure the actual movement (i.e., displacement) of a machine's shaft relative to the probe. A displacement probe must be rigidly mounted to a stationary structure to gain accurate, repeatable data.

Permanently mounted displacement probes provide the most accurate data on machines with a low rotor weight (i.e., low relative to the casing and support structure). Turbines, large process compressors, and other plant equipment should have displacement transducers permanently mounted at key measurement locations to acquire data for the program.

The useful frequency range for displacement probes is from 10 to

1000 Hertz, or 600 to 60,000 RPM. Frequency components below or above this range are distorted and, therefore, unreliable for determining machine condition.

The major limitation with displacement or proximity probes is cost. The typical cost for installing a single probe, including a power supply, signal conditioning, and so forth, averages $1000. If each machine in your program requires ten measurements, the cost per machine will be about $10,000. Using displacement transducers for all plant machinery would dramatically increase the initial cost of the program.

Displacement data are normally recorded in terms of mils (.001 inch), peak-to-peak. This value expresses the maximum deflection or displacement off of the true centerline of a machine's shaft.

Velocity Transducers

Velocity transducers are electromechanical sensors designed to monitor casing or relative vibration. Unlike the displacement probe, velocity transducers measure the rate of displacement — not actual movement. Velocity data are normally expressed in terms of ips (inches per second), peak and are perhaps the best way of expressing the energy created by machine vibration.

Velocity transducers, like displacement probes, have an effective frequency range of about 10 to 1000 Hertz. They should not be used to monitor frequencies below or above this range.

The major limitation of velocity transducers is their sensitivity to mechanical and thermal damage. Normal plant use can cause a loss of calibration; therefore, a strict recalibration program must be used to prevent distortion of data. At a minimum, velocity transducers should be recalibrated at least every six months. Even with periodic recalibration, programs using velocity transducers are prone to bad or distorted data that results from loss of calibration.

Accelerometers

Accelerometers use a piezeoelectric crystal to convert mechanical energy into electrical signals. Data acquired with this type of transducer indicate relative vibration, not actual displacement, and are expressed in terms of "*g's*" or inches/second/second. Acceleration is perhaps the best method of determining the force created by machine vibration.

Accelerometers are susceptible to thermal damage. If sufficient heat is allowed to radiate into the crystal, it can be damaged or destroyed. However, because the data-acquisition time, using temporary mounting techniques, is relatively short (i.e., less than 30 seconds), thermal damage is rare. Accelerometers do not require a recalibration program to insure accuracy.

The effective range of general purpose accelerometers is from about 1 Hertz to 10,000 Hertz. Ultrasonic accelerometers are available for frequencies up to 1 MHertz.

Machine data above 1000 Hertz (60,000 RPM) should be taken and analyzed in terms of acceleration ($g's$).

Mounting Techniques

Predictive maintenance programs using vibration analysis must have accurate, repeatable data to determine the operating condition of plant machinery. In addition to the transducer, three factors affect data quality: measurement point, orientation, and compressive load.

Key measurement point locations and orientation to the machine's shaft are selected as part of the data-base setup to provide the best possible detection of incipient machine-train problems. Deviation from the exact point or orientation will affect the accuracy of acquired data. Therefore, it is important that every measurement throughout the life of the program be acquired at exactly the same point and orientation. In addition, the compressive load (i.e., the downward force applied to the transducer) should be exactly the same for each measurement. For accuracy of data, a direct mechanical link to the machine's casing or bearing cap is absolutely necessary. Slight deviations in this load will induce errors in the amplitude of vibration and may also create false frequency components that have nothing to do with the machine.

The best method of ensuring that these three factors are exactly the same each time is to hard-mount vibration transducers to the selected measurement points. This guarantees accuracy and repeatability of the acquired data. It also increases the initial cost of the program. The average cost of installing a general purpose accelerometer will be about $300 per measurement point, or $3000 for a typical machine-train.

To eliminate the capital cost associated with permanently mounting transducers, a well-designed quick-disconnect mounting can be used. This mounting technique permanently mounts a quick-disconnect stud, with an average cost of less than $5, at each measurement point location. A mating sleeve built into a general-purpose accelerometer is then used to acquire accurate, repeatable data. A well-designed quick-disconnect mounting technique can provide the same accuracy and repeatability as the permanent mounting technique but at a much lower cost.

The third mounting technique that can be utilized is a magnetic mount. For general-purpose use below 1000 Hertz, a transducer can be mounted in conjunction with a magnetic base. Because the transducer/magnet assembly has a resonant frequency that may provide some distortion to the acquired data, the use of this technique would allow only marginal success. Because the magnet can be placed anywhere on the machine, there is no guarantee that the exact location and orientation would be maintained for each measurement.

The final method, used by some plants, to acquire vibration data is a hand-held transducer. This approach is not recommended if any other method can be used. Hand-held transducers do not provide the accuracy and repeatability required to gain maximum benefit from a predictive maintenance program. If this technique must be used, extreme care should be exercised to ensure that the same exact point, orientation, and compressive load are used for every measurement point.

GETTING STARTED

The steps defined above have been provided as guidelines for establishing a predictive maintenance data base. The only steps remaining to get the program started are to establish measurement routes and take the initial (or baseline) measurements. Remember, the predictive maintenance system will need multiple data sets to develop trends on each machine. With this data base, you will be able to monitor the critical machinery in your plant for degradation and begin to achieve the benefits that predictive maintenance can provide. The actual steps required to implement a data base are dependent on the specific predictive maintenance system selected for your program. The system's vendor should provide the training and technical support required to properly develop the data base with the information developed in the preceding chapters.

TRAINING

Successful completion of the critical phase of creating a total plant predictive maintenance program requires a firm knowledge of the operating dynamics of plant machinery, systems, and equipment. Normally, some (if not all) of this knowledge exists within the plant staff. However, the knowledge may not be held by the staff members selected to implement and maintain the predictive maintenance program.

In addition, a good working knowledge of the predictive maintenance techniques and systems included in the program is absolutely necessary. This knowledge (in all probability) is not currently possessed by your existing plant staff. Therefore, training is strongly recommended before establishment of your program. The minimum recommended level of training includes: user's training for each predictive maintenance system that will be used, a course on machine dynamics, and a basic theory course on each of the techniques (e.g., vibration, infrared) that will be used.

In some cases, all of these courses can be provided by the systems vendors. If not, there a number of companies and professional organizations that offer courses on most nondestructive testing techniques.

TECHNICAL SUPPORT

The labor and knowledge requirements for the proper establishment of a predictive maintenance program are often too much for plant staffs to handle. To overcome this problem, the initial responsibility for creating a viable, total plant program can be contracted to a company that specializes in predictive maintenance.

Some companies that provide full consulting and engineering services directed specifically toward predictive maintenance have the knowledge required and years of experience. These can provide all of the labor required to fully implement a total plant program and often can reduce the total time required to get the program up and running.

Caution should be used in selecting a contractor to provide this start-up service. Check references very carefully.

Chapter 9

Maintaining the Program

The labor-intensive part of your predictive maintenance management is complete. A viable program has been established, the initial data base is complete, and you have begun to monitor the operating condition of your critical plant equipment. Now what?

Most programs stop right here. The predictive maintenance team members do not continue their efforts to get the maximum benefits that predictive maintenance can provide. Instead, they rely on trending, comparative analysis, or (in the case of vibration-based programs) simplified signature analysis to maintain the operating condition of their plant. This is not enough to gain the maximum benefits from a predictive maintenance program.

In this chapter, methods are discussed that can be used to ensure maximum benefits from your program and, at the same time, improve the probability that the program will continue.

TRENDING TECHNIQUES

The data base that was established in Chapter 5 included broadband, narrowband, and full signature vibration data. It also included process parameters, bearing-cap temperatures, lubricating oil analysis, thermal imaging, and other critical monitoring parameters. What do you do with this data?

The first method required to monitor the operating condition of plant equipment is to trend their relative condition over time. Most of the microprocessor-based systems provide the means of automatically storing and recalling vibration and process parameters trend data for analysis. They can automatically prepare and print reports that quantify the operating condition at a specific point in time. A few can automatically print trend reports that quantify the change within a selected time frame. All of this is great, but what does it mean?

Monitoring the trends of a machine-train or process system enables the maintenance team to prevent most catastrophic failures. These trends are similar to the bathtub curve used to schedule preventive maintenance (Figure 1-1). The difference between the preventive and predictive bathtub curve is that the latter is based on the actual condition of the equipment, not on a statistical average.

The disadvantage of relying on trending as the only means of maintaining a predictive maintenance program is that it cannot tell you the reason a machine is degrading. One good example of this weakness is an aluminum foundry that relied strictly on trending to maintain its predictive maintenance program. In the foundry were 36 cantilevered fans that were critical to plant operation. The rolling-element bearings in each of these fans were changed on an average of every six months. By monitoring the trends provided by their predictive maintenance program, the staff was able to adjust the bearing change-out schedule based on the actual condition of the bearings in each fan. Over a two-year period, there were no catastrophic failures or loss of production that resulted from the fans being out of service.

Did this predictive maintenance program work? In their terms, the program was a total success. However, the normal bearing life should have been much greater than six months. Something in the fan or process created the reduction in average bearing life. Limiting their program to trending only, they were unable to identify the root-cause of the premature bearing failure. Properly used, a predictive maintenance program can identify the specific root-cause of chronic maintenance problems.

In the above example, an eventual full analysis provided the answer. Plate-out (material buildup) on the fan blades constantly increased the rotor mass, eventually forcing the fans to operate at critical speed. The imbalance created by operating at critical speed was the forcing function that destroyed the bearings. After taking corrective actions, the plant now gets an average of three years from the fan bearings.

ANALYSIS TECHNIQUES

All machines have a finite number of failure modes. If you have a thorough understanding of these failure modes and the dynamics of the specific machines, you can learn the vibration analysis techniques that can isolate the specific failure mode or root-cause of each machine-train problem.

The following example provides a comparison of various trending and analysis techniques.

Broadband Analysis

Each data point acquired using broadband techniques is a value that represents the total energy generated by the machine-train at the specific measurement point location, in the direction opposite the transducer. Most programs trend or compare the recorded value at a single point and disregard the other measurement points on the common shaft.

Rather than evaluate each measurement point separately, plot the energy of each measurement point on a common shaft. Figure 9-1

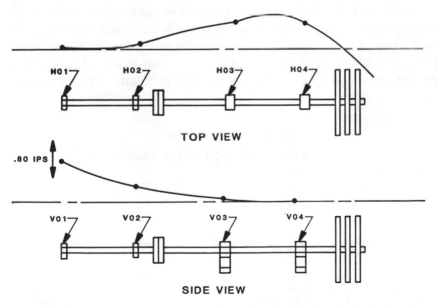

Figure 9-1. Plots of shaft mode shape. Plotting the broadband values in the horizontal and vertical plane will indicate the mode shape of the shaft during operation.

illustrates this technique for the Spencer blower data of Table 9-1. First, the vertical measurements were plotted to determine the mode shape of the machine's shaft. This plot indicates that the outboard end of the motor shaft is displaced, in the vertical direction, much more than the remaining shaft.

This limits the machine problem to the rear of the motor. Based strictly on the overall value, the probable cause is loose motor mounts on the rear motor feet. The second step involved plotting the horizontal mode shape. This plot indicates that the shaft is deflected, horizontally, between the pillow block bearings. Without additional information, the mode shaft suggest a bent shaft between the bearings. Even though the absolute failure mode cannot be identified, the trouble can be isolated to the section of the machine-train between the pillow block bearings.

Narrowband Analysis

The addition of unique narrowbands that monitor specific machine components or failure modes provides more diagnostic information.

Analysis of the narrowband information acquired from the Hoffman blower (Table 9-2) indicates that the vertical data are primarily at the true running speed of the common shaft. This confirms that a deflection of the shaft exists, and that no other machine component or failure mode is contributing to the problem. The horizontal

Table 9-1. Broadband

BROAD BAND RMS (IPS-RMS)	
V01	.80
V02	.30
V03	.10
V04	.08
H01	.03
H02	.10
H03	.50
H04	.50
A04	.25

Table 9-2. Narrowband RMS (IPS-RMS)

POINT	BROADBAND	NARROWBANDS		
		RUNNING SPEED 1X	2X	BLADE-PASS
V01	.80	.75	.05	.05
V02	.30	.25	.05	.05
V03	.10	.03	.01	.05
V04	.08	.02	.01	.06
H01	.03	.01	.01	.005
H02	.10	.01	.01	.05
H03	.50	.10	.30	.10
H04	.50	.10	.30	.10
A04	.25	.10	.03	.15

measurements indicate that the bladepass, bearing defect, and mis-alignment narrowbands are the major contributors.

As discussed above, fans and blowers are prone to aerodynamic instability. The indication of abnormal vane-pass suggests that this may be contributing to the problem. The additional data provided by the narrowband readings help to eliminate many of the possible failure modes that could be affecting the blower, yet the data cannot pinpoint the specific problem.

Root-Cause Failure Analysis

A visual inspection of the blower indicated that the discharge was horizontal and opposite the measurement point location. After a check of the process parameters recorded concurrently with the vibration measurements, the motor was found to be in a no-load or run-out condition, and the discharge pressure was abnormally low. In addition, the visual inspection showed that the blower sat on a cork pad and was not bolted to the floor. The discharge piping, 24-inch-diameter schedule-40 pipe, was not isolated from the blower nor did it have any pipe supports for the first 30 feet of horizontal run.

With all of these clues in hand, the examiners concluded that the blower was operating in a run-out condition (i.e., it was not generating any pressure) and was, therefore, unstable. This part of the machine problem was corrected by reducing (i.e., partially closing) the damper setting and forcing the blower to operate within acceptable aerodynamic limits. After correcting the damper setting, all of the abnormal horizontal readings came within acceptable limits.

The vertical problem with the motor was isolated to improper installation. The weight of approximately 30 feet of discharge piping compressed the cork pad under the blower and forced the outboard end of the motor to elevate above the normal centerline. In this position, the motor became an unsupported beam and resonated in the same manner as a tuning fork. After the discharge piping from the blower was isolated and support provided, the vertical problem was eliminated.

If you follow the suggested steps in Chapter 5, your predictive maintenance team will receive training on how to use the predictive maintenance system or systems selected for your program. In addition, they will be exposed to the theory behind each of the techniques that will be used to analyze the data acquired by the systems. Will this be enough to gain maximum benefit from your program?

ADDITIONAL TRAINING

The initial user's training and basic theory will not be enough to gain maximum benefits from a total plant predictive maintenance program. You will need to continue the training process throughout the life of the program.

There are a variety of organizations, including predictive maintenance systems vendors, that provide training programs in all of the predictive maintenance techniques. Caution in selecting both the type of course and instructor is strongly recommended. Most of the public courses are, in reality, sales presentations. They have little practical value and will not provide the knowledge base required to gain maximum benefit from your program.

Practical or application-oriented courses are available that will provide the additional training required to gain maximum diagnostic benefits from your program. The best way to separate the good from the bad is to ask previous attendees. Request a list of recent attendees and then talk to them. If the courses are presented by reputable firms, they will gladly provide this information.

TECHNICAL SUPPORT

None of the predictive maintenance technologies are capable of resolving every possible problem that may develop in a manufacturing or process plant. For example, the microprocessor-based vibration systems use single-channel data collectors. These systems cannot monitor transient problems, torsional problems, and many other mechanical failures that could occur. At best, they can resolve 85% to 90% of the most common problems that will occur.

To resolve the other 10% to 15% of mechanical problems and to conduct the other nondestructive testing that may occasionally be required to maintain the plant, you will need technical support. Few of the predictive maintenance systems vendors can provide the level of support required. Therefore, you will need to establish contacts with consulting and engineering services companies that have a proven record of success in each of the areas required to support your program.

There are consulting and engineering services companies that offer full support to predictive maintenance. These companies specialize in the nondestructive testing and analysis techniques required to solve plant problems. Caution in selecting a technical support contractor is recommended. As in training suppliers, there are ten bad ones for every good one.

Contract Predictive Maintenance Programs

The benefits that are derived from a total plant predictive maintenance program provide the means of controlling maintenance costs, improving plant performance, and increasing the profit of most manufacturing and production plants. Unfortunately, many plants do not have the staff to implement and maintain the regular monitoring and analysis that is required to achieve these goals. There is a solution to this problem.

The proven benefits derived from predictive maintenance and the staff limitations at numerous plants have created a new type of service company. There are numerous reputable companies that now specialize in providing full capability predictive maintenance services on an annual contract basis. These companies will provide all of the instrumentation, data-base development, data-acquisition and analysis responsibility, and periodic reports that quantify plant condition. Using contract predictive maintenance will provide plants with all of the benefits of predictive maintenance without the major expense required to set up and maintain an in-house program.

As stated, there are many reputable companies that can provide this service. However, there are also a great many that claim to provide full predictive maintenance services and do not. Extreme caution must be exercised in the selection process. As in the case of selecting a system and vendor for an in-house program, references should be thoroughly checked.

Chapter 10

Characteristics of Vibration

All machines generate unique mechanical forces as part of their normal operation. As the mechanical condition of the machine changes, so do the forces. Understanding the dynamics of operating machinery and how the forces create unique vibration-frequency components is the key to using frequency-domain vibration as a predictive maintenance tool. This chapter provides a basic understanding of vibration terms.

VIBRATION THEORY

Vibration in its general sense is periodic motion. In other words, the motion will repeat itself in all its particulars after a certain interval of time. The simplest kind of periodic motion is a harmonic motion; in it, the relation between the maximum displacement and time may be expressed by:

$$X = X_0 \sin \omega t.$$

The maximum value of the displacement is X_0, called the amplitude of the vibration. The period T usually is measured in seconds. Its reciprocal, $f = 1/T$, is the frequency of the vibration, measured in cycles/second (CPS) or Hertz (Hz). The symbol ω, which is known as the circular frequency, is measured in radians/second. From Figure 10-1 it is clear that a full cycle of the vibration takes place when ωt has passed through 360 degrees, or 2π radians; then the sine function resumes its previous values.

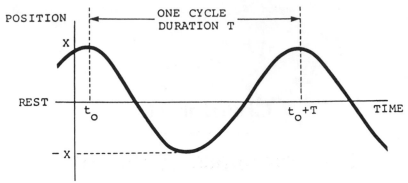

Figure 10-1. A plot of simple vibration.

For rotating machinery, the frequency is often expressed in vibra-tions/minute (VPM): VPM $= 30 \ \omega/\pi$. In a harmonic motion for which the displacement is given by $X = X_0 \sin \omega t$, the velocity is found by differentiating the displacement with respect to time,

$$\frac{dX}{dt} = \dot{X} = X_0 \omega * \cos \omega t,$$

so that the velocity is also harmonic and has a maximum value ωX_0. The acceleration of motion is:

$$\frac{d^2X}{dt^2} = \ddot{X} = -X_0 \omega^2 \sin \omega t$$

and is also harmonic with a maximum value $\omega^2 X_0$. Consider two vibrations given by the expressions $X_1 = a \sin \omega t$ and $X_2 = b \sin (\omega t + \phi)$, which are shown in Figure 10-2, plotted against ωt as the abscissa. Owing to the presence of the quantity ϕ, the two vibrations do not attain their maximum displacements at the same time, but the one is $\phi \omega$ seconds behind the other. The quantity ϕ is known as the phase angle, or phase difference, between the two vibrations. From the graph, it is seen that the two motions have the same ω (circular frequency) and, consequently, the same frequency. A phase angle has meaning only for two motions of the same frequency: If the frequencies are different, phase angle is meaningless.

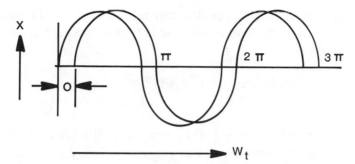

Figure 10-2. Two vibrations which exhibit a phase difference.

All harmonic motions are periodic; but not every periodic motion is harmonic. For example, Figure 10-3 represents a nonharmonic motion of two sine waves where

$$X = (a)\sin \omega t + (a/2)\sin 2\omega t$$

This superposition of two sine waves of different frequencies is a periodic motion but not harmonic.

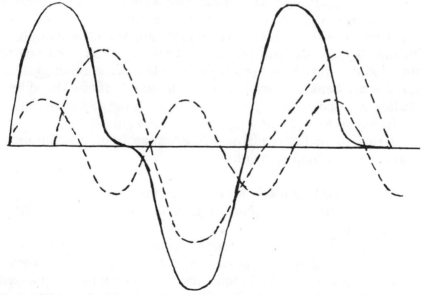

Figure 10-3. The sum of two harmonic motions having different frequencies is not a harmonic motion.

Mathematical theory shows that any periodic curve of frequency ω can be split up into a series of sine curves of frequencies ω, 2ω, 3ω, and so forth:

$$f(t) = A_O + A_1\sin(\omega t + \phi_1) + A_2\sin(2\omega t + \phi_2) +$$
$$A_3 \sin (3\omega t + \phi_3) + \quad \ldots$$

provided that the frequency $[\, f(t)\,]$ repeats itself after each interval $T = 2\pi/\omega$. The amplitudes of the various waves (e.g., A_1, A_2) and their phase angles (e.g., ϕ_1, ϕ_2) can be determined analytically when $f(t)$ is given. This calculation is known as a Fourier series. The second term (e.g., 2ω, 3ω) indicates the harmonics of the primary frequency. In most vibration signatures, the primary frequency component will be at one of the true running speeds of the machine-train (1X or 1ω), but will also have one or more harmonics at 2X, 3X, and other harmonics of the primary running speed. It is not necessary to get deeply involved in the mathematics of vibration other than to know that, by using sound engineering analysis techniques, the vibration components, amplitudes, and phase can be quantified.

In theory, a perfectly balanced machine without any forcing functions (e.g., gravity, friction) would be totally free of any vibration. In the practical world, all machinery and machine-trains have multiple vibration frequencies as a part of normal operation. The key to using vibration analysis for predictive maintenance is understanding the difference between normal and abnormal vibration. This differentiation is based primarily on two values: (1) the vibration amplitude (strength) and (2) the frequency component.

Two rules of predictive maintenance, using vibration monitoring, should be committed to memory:

1. There *must* be a mechanical and/or process-related reason for every frequency component in a machine-train's vibration signature.
2. The amplitude of each frequency component does not change without a reason. Any deviation of the frequency components and/or amplitude is an indication that something has changed in the operating dynamics of that specific machine-train.

Please note that vibration amplitude may decrease or increase. Either change may indicate degradation of the machine-train. Do not assume that a lower amplitude indicates an improvement in the mechanical condition of the machine-train.

VIBRATION TERMINOLOGY

The analysis of vibration requires an understanding of the terminology used to describe the components of vibration. The following sections define the terms that are normally used in vibration analysis:

1. Frequency
2. Amplitude
3. Displacement
4. Velocity
5. Acceleration
6. Time Domain
7. Frequency Domain
8. Broadband Analysis
9. Narrowband Analysis
10. Signature Analysis

Frequency

The cyclic movement occurring in a given unit of time, frequency is typically expressed as (1) RPM, revolutions per minute, or (2) CPM, cycles per minute. Frequency is most commonly expressed in multiples of the rotation (running) speed of the machine-train. This is primarily due to the tendency of machine-train vibration frequencies to occur in direct multiples or submultiples of the rotating speed of the machine. It also provides an easy means to express the frequency of vibration. It is simpler to refer to the frequency of vibration as one times RPM (1X), two times RPM (2X), and so forth, rather than having to express all vibration in cycles per minute or Hertz (Figure 10-4). Some malfunctions tend to occur at specific frequencies; this fact helps to segregate certain classes of malfunctions from others.

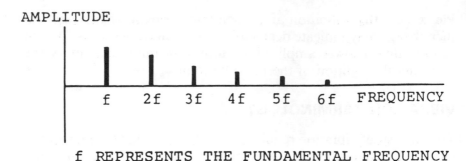

f REPRESENTS THE FUNDAMENTAL FREQUENCY

Figure 10-4. The discrete frequency spectrum of a sawtooth wave.

It is important to note, however, that the frequency-malfunction relationship is not mutually exclusive. A specific mechanical problem cannot definitely be attributed to a unique frequency. Frequency is a very important piece of information with regard to analyzing rotating machinery, and will help to isolate the malfunctions, but it is only one part of the total. It is necessary to evaluate all data before arriving at a conclusion.

Amplitude

The magnitude of dynamic motion or vibration amplitude is typically expressed in terms of (1) Peak-to-Peak, (2) Zero-to-Peak (peak), or (3) RMS (Root-Mean-Square). Figure 10-5 illustrates the relationship of the three units of measure associated with amplitude. Amplitude—whether expressed in displacement, velocity, or acceleration—is an indicator of severity. Since industrial standards of vibration severity are usually expressed in one of these terms, you should have a clear understanding of their relationship. Care must be exercised to note the "type" of amplitude measurement when comparing machine-train vibration to standards.

Displacement

The change in distance or in the position of an object relative to a reference point, displacement is usually expressed as mils (0.001 inch). Displacement is the actual distance off the true centerline of a

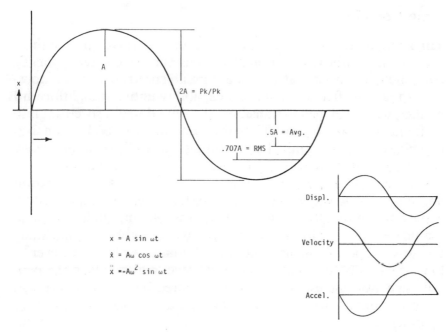

$$x = A \sin \omega t$$
$$\dot{x} = A\omega \cos \omega t$$
$$\ddot{x} = -A\omega^2 \sin \omega t$$

Figure 10-5. The relationships of displacement, velocity, and acceleration in vibration measurements.

rotating shaft as compared to a stationary reference (usually the machine housing). In simple terms, displacement is the actual radial or axial movement of the shaft in relation to the normal centerline.

Velocity

The time rate of change of displacement or position, velocity is usually expressed as ips, inches per second. It is the rate (frequency) of movement (displacement). Simply stated, velocity is the number of times per unit time that displacement takes place.

Acceleration

The time rate of change of velocity, acceleration is typically expressed in inches/second/second, or g's ($1g = 32.17$ ft/sec/sec).

Time Domain

Time domain provides a vibration signature for a given unit of time. Most of the early vibration analysis was carried out using analog equipment and necessitated a time-domain spectrum. It was impossible to provide frequency domain capability until a straightforward method was developed to transform the time-domain spectrum into its frequency components. Time-domain data are good for analyzing phase, but are limited in defining the individual frequency components that contribute to a machine-train's signature.

Data collection using time domain is typically single spectrum and may tend to show nonrecurring peaks, or spikes (Figure 10-6). Time domain usually provides only an overall amplitude value and the phase relationship. It is sometimes useful to observe and study the waveshape of signals in a time-domain display. However, a low-signal amplitude and a high-noise level may result in the periodic signal of interest being totally obscured by the random noise.

Time domain has the following characteristics that must be considered:

1. A reference, or trigger, synchronizing pulse must be provided as part of the data-acquisition system. The trigger time-tags the data and sets the duration of acquisition.
2. The actual waveshape of the signal is recovered.
3. Transient waveshapes may be recovered as long as a coherent reference signal is available.
4. The phase, or time, relationship between two signals can be measured.

This process is very powerful when a periodic signal at one frequency is buried in noise and a synchronizing pulse is available. The disadvantages become apparent when one considers the recovery of a blade-passing signal in a gas turbine compressor. If the compressor contained only one stage, then this technique could increase the signal-to-noise ratio such that you could observe the waveform generated by each blade passing. However, if there are many stages, all the blade-passing signals will be synchronized to the same shaft speed. While repeated signal averaging will increase the signal-to-noise ratio, the "signal" will be the sum of all blade-passings. If

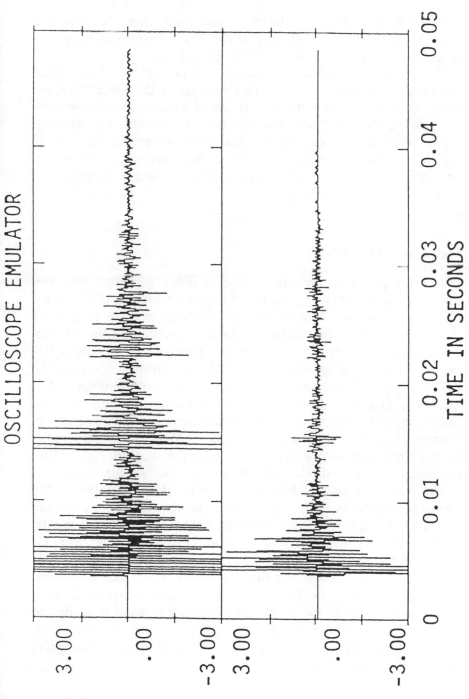

Figure 10-6. The single spectrum time domain signature shows nonrecurring peaks.

filtering is used ahead of the microprocessor, then the observed waveshape will be distorted because the higher harmonics necessary to define the true waveshape will be filtered out.

In summary, time-domain analysis is a powerful tool in special circumstances. However, its value decreases as the vibration spectrum becomes a more complex mixture of components all synchronized to the same shaft speed or where a component is a function of the interaction between two shafts at different speeds. Since most typical machine-trains have complex, multiple-vibration-component spectra, the use of time domain in a predictive maintenance program is severely limited.

Frequency Domain

Frequency domain converts a time domain spectrum into frequency components using the Fast Fourier Transform (FFT). Simply stated, frequency domain converts the time-based spectrum to a true representation of the individual frequency components. Because most machine-train failures are displayed at or near a frequency component associated with the true running speed(s), the ability to display and analyze the spectrum as components of frequency is extremely important. With frequency-domain analysis, the average spectrum for a machine-train signature can be obtained. Nonrecurring peaks, or spikes, can be normalized to present an accurate representation of the machine-train condition. Figure 10-7 illustrates a simplified relationship between the two methods.

The real advantage of frequency-domain (FFT) analysis is the ability to normalize each vibration component so that a complex machine-train spectrum can be divided into discrete components. This ability simplifies analysis and isolation of mechanical degradation with the machine-train. In addition, it should be noted that frequency-domain analysis is the only method that can determine the phase relationships for all of the vibration components in a typical machine-train spectrum. Time-domain analysis is limited to one running speed, whereas frequency domain can normalize any or all running speeds. Thus, phase relationships of all first-order (1X) speeds can be determined.

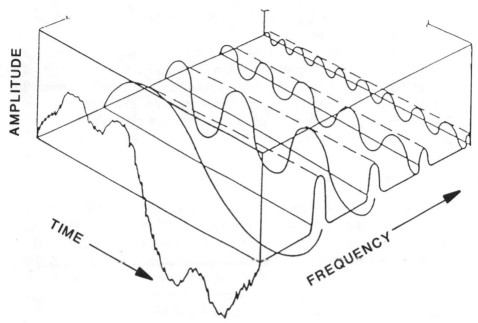

Figure 10-7. The relationship between time- and frequency-domain signatures.

Broadband Analysis

Broadband analysis techniques have been used for monitoring the overall mechanical condition of machinery for over 20 years. The technique acquires the overall vibration (energy) from zero to the user-selected maximum frequency (F_{max}). Broadband data is strictly a gross overall vibration expressed in RMS (displacement, velocity, or acceleration) energy and does not provide any indication of the specific frequency components (e.g., 1X, 2X) that make up the frequency-domain signature of the machine-train's vibration. As a result, identification of specific machine-train problems cannot be isolated or identified.

The lack of actual frequency-domain (FFT) signatures diminishes the value of the broadband data. Since broadband data are limited to the gross overall energy of all frequency components, and do not provide specific information for each of the unique frequency components, little actual analysis of specific problems within the machine-train is possible.

The only useful function of broadband analysis is long-term trending of the gross overall condition of machinery. Typically, a set of alert/alarm limits is established to monitor and trend the overall condition of the machine-trains in a predictive maintenance program. This approach has limited value and severely restricts the full power of a comprehensive program.

Narrowband Analysis

Narrowband, like broadband analysis, monitors the overall energy (RMS) of a band of frequency components. However, the ability to select specific groups of frequencies (narrowbands) increases the value of the data. Unlike broadband analysis, narrowband provides the ability to directly monitor, trend, and alarm specific machine-train components. With narrowband, a window of frequencies unique to specific machine components can be established, directly monitored, and trended automatically by the microprocessor. For example a narrowband window can be established to directly monitor a gear-set energy that consist of the primary gear-mesh frequency and corresponding sidebands.

Using this technique, a set of alert/alarm limits can be established for each of the narrowbands. Even though the data are limited to the overall energy (RMS) of each narrowband window, this technique provides the means to automatically monitor and alarm specific machine components.

The narrowband technique drastically reduces the amount of staff required to monitor machine-trains and improves the accuracy of detecting incipient problems.

As in the case of broadband analysis, narrowband does not provide spectral display of the specific frequency components within the window. Therefore, problem diagnostics is limited to the overall (RMS) energy.

Signature Analysis

This term is usually applied to the vibration spectrum which uniquely identifies a machine, component, system, or subsystem at a specific operating condition and point in time. A full signature (FFT) provides specific data on every frequency component within the overall frequency range of a machine-train (Figure 10-8). The typical frequency range can be from 0.1 Hertz to 20,000 Hertz or more.

In microprocessor systems, the signature is formed by breaking the total frequency spectrum down into unique components (peaks). Each line or peak represents a specific frequency component that, in turn, represents one or more mechanical components within the machine-train. Typical microprocessor-based predictive maintenance systems can provide signature resolutions of 100, 200, 400, 800, and 1600 lines.

Full-signature spectra are important analysis tools, but they require a tremendous amount of microprocessor memory. It is impractical to collect full spectra on all machine-trains on a routine

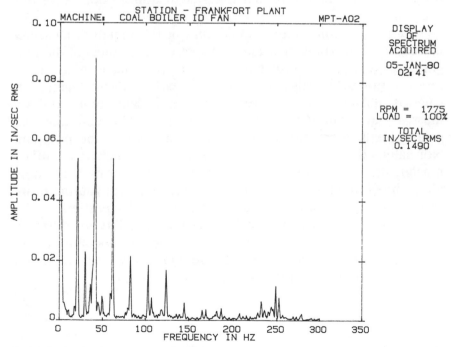

Figure 10-8. A full signature provides specific data on every frequency component within the overall frequency range of a machine train.

basis. A typical predictive maintenance microprocessor can only store 1–200 full signatures, or 10 to 20 machine-trains at one time. In addition, the data management and storage in the host computer would be extremely difficult and costly.

Full-range signatures should be collected only if a confirmed problem has been identified on a specific machine-train. Typically, these are automatically triggered by exceeding a preset alarm limit or obtained on demand (manual input) during the data-acquisition operation.

Typically, a machine-train's vibration signature is made up of vibration components, with each component associated with one or more of the true running speeds within the machine-train. Most machinery problems will show up at or near one or more of the true running speeds. Therefore, the narrowband capability is very beneficial in that high-resolution windows can be preset to monitor the running speeds. Many of the microprocessor-based predictive maintenance systems available do not have narrowband capability. Care should be taken to ensure that the system utilized does have this capability.

Systems that utilize either broadband or full-signature measurements have limitations that may hamper the usefulness of the program. Broadband measurements typically do not have enough resolution at running speeds to be effective in early problem diagnostics, whereas full-signature measurements at each data point require a massive data-acquisition, handling, and storage system that will greatly increase the capital and operating costs of the program. Normally, full-signature spectra are needed only when an identified machine-train problem demands further investigation. Please note that, while full signatures provide too much data for routine problem detection, they are essential for root-cause diagnostics.

The optimum system would include the capability to utilize all three techniques. This ability would optimize the program's ability to trend and do full root-cause failure analysis, yet still maintain minimum data-management and storage requirements.

Chapter 11

Vibration Severity Criteria

Industrial standards for vibration severity are typically divided into casing vibration and shaft vibration. The following sections explain and quantify both values.

CASING VIBRATION

Although it would be convenient to think of vibration limits as exact, concrete values, they generally are not. As soon as someone says all machines will operate satisfactorily below a certain level or will not operate above a certain level, exceptions will be cited. Thus, assessing vibration severity is largely a matter of experience aided by other guidelines (e.g., machine-train vibration history).

Vibration amplitude severity charts had their beginning in 1939, when T. C. Rathbone published recommended limits based on his experience as an insurance inspector. These Rathbone limits were based on casing measurements taken on heavy slow-speed machines with a 2:1 or 3:1 ratio between shaft vibration and bearing-housing vibration. Rathbone's idea of plotting maximum tolerable-vibration levels versus frequency was refined over the years by a number of people and finally evolved into the chart shown in Figure 11-1. One feature of this and all vibration severity charts should be immediately apparent: In terms of displacement, tolerable vibration amplitudes decrease with increasing frequency.

The severity chart presented in Figure 11-1 and most similar vibration severity charts based on constant velocity are designed for casing measurements taken on typical machines with casing-rotor

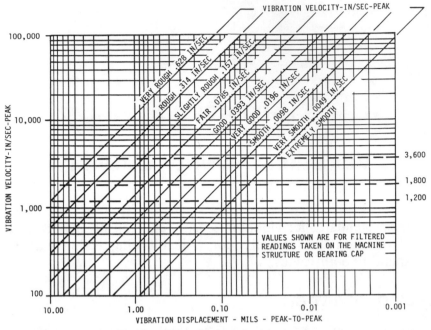

Figure 11-1. Vibration severity charts are based on constant velocity (speed) and are designed for readings taken from the machine casing or bearing caps.

weight ratios on the order of 5:1. Before using any limit, one must know what the limit applies to and how the measurement must be made. Failure to observe these precautions can lead to erroneous conclusions. Two standards for judging vibration severity were published by the ISO. Widely used in Europe, ISO Standards 2372 and 3945 are, respectively, a general standard designed primarily for shop testing and acceptance, and a more specific standard designed for evaluating the vibration of larger machinery *in situ*. Both standards contain criteria for judging machine condition for casing velocity measured at a specified location at each bearing. The standards apply to machines operating within the speed range from 10 to 200 Hertz (600 to 12,000 RPM.) They specify a measurement limited to a frequency band of 10 to 1000 Hertz.

These two standards require a true RMS amplitude measurement and specify that the more common, rectified value scaled to read RMS with a sine wave is not acceptable unless the vibration is a pure sinusoid. This is probably one of the most controversial aspects of the ISO casing velocity standard; many believe that a peak amplitude is a better measure of severity than an RMS value. Favoring RMS amplitude measurement is the fact that RMS amplitude is more closely related to the energy content of the signal. A signal can have a high peak value without much energy content if the duration (width) of the peak is small.

Both standards make a distinction between flexible-support and rigid-support machines and recognize that a support system may be rigid in one plane and flexible in the other. As in rotor dynamics, a flexible support is defined as a support having its first natural frequency below the main frequency of excitation, presumably machine running speed.

Conversely, a rigid support is one for which the first natural frequency of the support structure is higher than the main excitation frequency. The ISO standards also recognize that machine casing vibrations can be transmitted from the environment and are not applicable when the transmitted excitation is greater than one-third of the operating value. Although several velocity limits are in use, there is good agreement between the various limits. Most machinery vibration analysts consider anything below 0.1 ips (Velocity) as a tolerable level of vibration, and anything above 0.6 ips-Peak (Velocity) as intolerable.

As indicated in Figure 8-1, acceptable vibration limits are speed dependent. This is always true with displacement measurements where one must know the frequency at which the amplitude occurs in order to assess its severity. While 4 mils would be tolerable at 1800 RPM, only about 0.6 mil at 10,000 RPM can be allowed. Table 11-1 represents a composite of the referenced data and is offered as a guideline for casing vibration severity:

Table 11-1. Recommended Limits for Overall Casing Vibration (Velocity)

PEAK VALUE (IPS)	QUALITY
Less than 0.15	Acceptable
0.15 to 0.25	Tolerable
0.25 to 0.40	Probably tolerable for moderate periods of time, but increase the monitoring frequency to warn of changes
0.40 to 0.60	Failure probable; watch changes closely and be prepared to shut down for repairs
Above 0.60	Danger of immediate failure

The condition under which these severity limits are applicable are limited-bandwidth casing measurements made on general-purpose industrial machinery. With atypical equipment, the principles are the same; all that must be done is to modify the severity criteria to fit the specific conditions. For example, it is often necessary to take casing measurements on vertically split compressors with casing-rotor weight ratios of 30:1 or higher, and bearings supported through 360 degrees. The impedance from shaft to casing is quite high on this type of equipment, with the result that casing vibration amplitudes will be very low. A reasonable approximation of the casing velocity limits that should be applied to vertically split machinery can be obtained by multiplying the values by 0.2. Sometimes the opposite is the case—the recommended limits are too low for a specific machine. Examples of this category are gearboxes and pumps that normally have high-frequency excitation created by gearmesh, vane-pass, and so forth.

Generally, multiplying the limits by 1.25 will result in more realistic limits for the equipment listed. Increasing limits should not, however, be accomplished without studied consideration. Even though these limits are empirical and can probably be exceeded for short periods without damage, there are no absolutes.

SHAFT VIBRATION

Vibration severity criteria for shaft displacement is somewhat more definitive than casing vibration. However, the vibration limits depend on several variables, such as measurement location and shaft mode. Figure 11-1 provides the best guidelines for shaft vibration. Here again, observe that tolerable displacement decreases with speed. A word of caution: the amplitudes shown on the chart are measured adjacent to a bearing and must be corrected for runout. Remember that the signal received from a noncontact displacement probe is a summation of actual shaft motion *and* any runout or defects on the shaft. As a result, the chart must be entered with the actual shaft motion.

Another point, the severity chart is generally entered with the machine running speed to determine a tolerable level of overall vibration. Generally this is a safe assumption, for in most cases running frequency will be the dominant component present in a

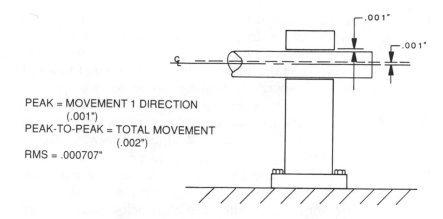

PEAK = MOVEMENT 1 DIRECTION
(.001")
PEAK-TO-PEAK = TOTAL MOVEMENT
(.002")
RMS = .000707"

*DISPLACEMENT IS THE **ACTUAL** DISTANCE
THAT A SHAFT MOVES OFF CENTERLINE
OR AXIALLY DOWN THE CENTERLINE*

Figure 11-2. Shaft vibration severity charts are based on the actual shaft displacement from it's centerline.

shaft vibration spectrum. Make sure that the true running speed of the proper shaft is used for each data point. Many machines have multiple running speeds, and it is easy to apply the wrong speed when evaluating vibration severity.

For example, if shaft motion is predominant at another frequency because of misalignment or instability, a limit based on running speed may be too high or too low for the particular conditions. To be more specific, there are numerous examples of high-speed machines which have been operated without any distress for extended intervals with subsynchronous instability present at levels far in excess of what would be tolerable if the excitation was at running speed. Vibration developed by the rotating shaft is attenuated by the impedance (resistance to motion) of the casing and the mounting structure.

Variations in impedance can cause large variations in the ratio between shaft and casing vibration (sometimes on the same machine). To cite a specific example, the outboard bearings on a large power-generating, low-pressure turbine are typically supported on concrete pedestals and are much stiffer than the inboard bearings, which may be supported on structural steel. With this type of design, it is not unusual to observe a ratio of shaft-to-casing vibration in the vicinity of 1:1 where the bearings are flexible, compared to a ratio on the order of 10:1 where the bearings are supported on a stiff pedestal. Here again, do not use these values as representative of a situation but consider the specific conditions existing at the location at which the measurement is made and adjust limits accordingly.

A word of caution concerning the use of any vibration limit: Although severity charts and limits are customarily divided into categories of machine condition, the progression is, of course, continuous. In other words, an amplitude slightly below a dividing line does not imply significantly better machine condition than one slightly above the same line. Industry organizations such as the American Petroleum Institute and the American Gear Manufacturers Association specify that, for the purpose of acceptance, maximum shaft amplitude in mils, peak-to-peak shall not exceed

$$\sqrt{\frac{12,000}{\text{RPM}}}$$

or 2 mils, whichever is less. In this criteria, displacement amplitude is measured shaft motion and includes runout, which can be no more than 0.25 times the allowed displacement. If the maximum acceptable amplitudes given by the preceding equation are plotted against speed, they fall well within the satisfactory range established. This conservative criterion is intended as an acceptance standard for new machinery and must be lower than what one might be willing to tolerate on operating equipment, where production is the primary objective.

Above 600 Hertz, the AGMA specification shifts to a constant $10g$ casing acceleration and establishes a criterion for evaluating high-frequency vibration related to gear quality and conditions at mesh. While most operating gears will meet this criterion, it appears a reasonable (if a bit high) acceptance criterion for new gears. Occasionally gears will operate successfully at amplitudes above $10g$.

There is a draft ISO standard covering shaft measurements. It specifies the measurement of shaft amplitude in peak-to-peak units across a frequency range from 10 to 1000 Hertz, from two transducers spaced at 90 degrees apart at each shaft bearing. Both absolute and relative measurements are covered. The standard also specifies that total runout will not exceed 10% of the allowable amplitude, a limit which is considerably stricter than the API runout limits of 25%. The draft standard includes judgement criteria for shaft displacement measured on large turbomachines, which are straight lines somewhat more restrictive than those shown in Figure 11-1.

RECOGNIZING CHANGES IN MACHINERY CONDITION

There is another factor involved in judging machine-train condition. A stable (unchanging) amplitude of any vibration component implies a stable mechanical condition. However, a trend that indicates either an upward or downward change in any vibration component is a sure indication of mechanical change. The amplitude of a specific vibration component will not change, either upward or downward, without a definable reason. Although this should be obvious, it is always worth repeating — for often the secret in avoiding a machine-train problem is early detection. For example, assume that a vibration survey is done on a machine-train that is about to be taken out of service for routine preventive maintenance. All vibration levels are within limits, so the decision is made not to do any work on the machine. Is this a good decision? It may not be, if any of the vibration components are trending upward at a rate that would exceed acceptable limits before the next scheduled shutdown. Thus, measurements alone are not enough to define mechanical condition. One must also know how the measurements are changing with time in order to reach a valid conclusion.

ESTABLISHING PRACTICAL EVALUATION CRITERIA

General limits are acceptable for a large percentage of operating machine-trains. There are, however, situations where it is highly desirable to refine the evaluation process and devise specific and perhaps more detailed criteria for a specific purpose or a specific machine-train. Thus far, the severity limits that have been discussed do not provide any information about what might be causing a problem, nor are they detailed enough to enable early recognition of small changes in mechanical condition, especially on complex equipment. In order to accomplish this necessary next step, a more discrete method of displaying vibration characteristics is needed, such as an amplitude versus frequency spectrum (frequency domain), and guidelines for evaluating the information must be developed.

The limiting values of vibration contained in Figure 11-1 assumes the signal is predominantly at the running frequency (speed) of the machine in question. A signal made up of a series of components, each at or close to the maximum allowable amplitude for that specific frequency, is clearly abnormal, even though the chart limits may not be violated. The question is how to evaluate a complex presentation containing numerous individual components? In low-frequency regions, use the limits presented in the severity charts to judge the running-speed component, and then compare the relative amplitudes of the remaining components to the running-speed component. To evaluate more complex spectra, the best method is a comparison with an established baseline for the specific machine-train and/or similar machine-trains.

For even more accuracy, compute statistical median values for each component in the frequency spectra generated by similar machine-trains, and establish a list of maximum and minimum values to be used as a performance envelope. The same results can be found with trending data taken over an extended time for the specific machine-train and/or similar machine-trains. The real key to this evaluation is accurate acquisition, storage, and trending of full-spectrum vibration data for each machine-train within the program. This process is relatively simple if a suitable microprocessor system is utilized.

Relative Strength of Vibration Components

A normal low-frequency vibration spectrum, and some abnormal spectra, will be dominated by running frequency. To gain an idea of machine condition apply the severity chart limits, modified for the specific machine type. If the amplitudes of the added components are small (less than about 1/3 the amplitude at running speed) and decreasing in amplitude with increasing frequency, the spectrum is normal and can be judged on the basis of the amplitude at running speed.

If the spectrum contains components either above or below running speed in excess of about 1/2 the amplitude at running speed, then some fault may be present and the spectrum will have to be evaluated for an assessment of the condition. Another clear abnormality is the presence of a nonsynchronous component above running speed.

There are a couple of exceptions to the foregoing general criteria which should be noted. The spectrum recorded from a gearbox normally contains components at shaft running frequencies which should be individually compared to severity limits. In fact, a gear spectrum might be seen as two discrete spectra superimposed on one another. Following this process, each running speed and its harmonics are evaluated exactly as described for spectra generated by machinery with only a single running speed. Gas turbines with two or more shafts fall into this same category; it is necessary to isolate and evaluate the characteristics produced by each shaft.

Finally, some spectral components are transmitted structurally from one machine-train to another where they may dominate the spectrum. It is not unusual for the casing spectrum of an outboard, gear-driven machine to contain a strong component at the running speed of the low-speed driver. To evaluate the condition of the high-speed unit, the presence of a low-speed component must be isolated, evaluated separately, and nulled from the high-speed spectrum before an accurate analysis can be made.

Chapter 12

Diagnostics and Analysis

There are two definitions of predictive maintenance used by industry. The first definition states that predictive maintenance provides the means to determine (1) if a problem exists in a machine-train, (2) how serious the problem is, and (3) how long the machine can run before failure. To meet this definition of predictive maintenance, little machine-train knowledge and/or vibration-analysis training is required. Many of the predictive maintenance systems that provide the ability to acquire and trend the RMS (energy) levels of machine-trains can provide all of the information needed to meet the definition's requirements.

The second definition of predictive maintenance is that the program should detect and identify specific machine components (e.g., gearsets, bearings) that are degrading. This definition requires that the user identify the failure mode that is causing the degradation. This type of analysis requires full-signature (FFT) data and a knowledge of both the operating mechanics/dynamics of machine-trains and vibration-analysis techniques.

ANALYSIS USING RMS DATA

RMS data, like time-domain data, provide an overall indication of the mechanical condition of a machine-train. Since detailed signatures (FFTs) are not available, it is extremely difficult to isolate the specific machine component or failure mode within the machine.

Two types of RMS data are typically provided by microprocessor-based predictive maintenance systems: (1) broadband (overall)

RMS and (2) narrowband RMS. Both provide the total energy within the frequency windows selected by the user. Neither provides amplitude of the specific frequency components that make up the window. Most microprocessor-based systems will automatically monitor, trend, and report changes in the RMS levels. Some will automatically alarm and project days-to-failure. Manual analysis is minimal. Unfortunately, so is the diagnostic power that frequency-domain vibration analysis provides.

Diagnostics is severely limited using only RMS data. However, you can determine if a problem exists. There are three methods that are generally used to determine the severity of a problem using RMS data: (1) comparison to baseline data, (2) comparison to industrial standards, and (3) trending the amplitude and rate of change.

Figure 12-1 illustrates the comparison of a current data set to the original baseline reading. This type of comparison provides a gross

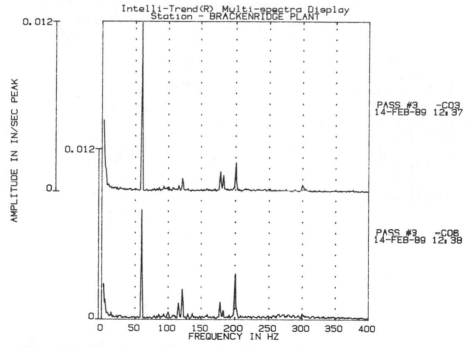

Figure 12-1. Comparison of vibration data. This type of comparison provides the gross relationship of two or more data sets.

comparison of relative energy which allows you to determine the overall mechanical condition. The original (baseline) RMS on February 14, 1989 was 0.3945 ips-RMS; a later reading on that date was 0.8398 ips-RMS. Monitoring the overall RMS, the rate of change, and the gross change in RMS energy will indicate if a problem is developing in the machine-train.

Figure 12-2 illustrates the same trend with alert/alarm limits, based on industrial standards, to show the severity of the overall vibration energy in the machine-train. The industrial standards are used as a guideline for determining the severity of the problem.

Neither of these trends provides any indication of the specific machine-train problem. The only information available to the user is an approximation of the relative mechanical condition of the machine.

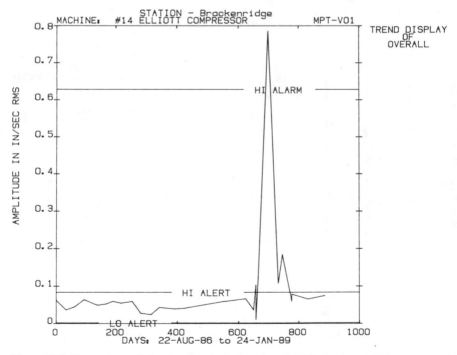

Figure 12-2. Comparison of vibration data including alert alarm limits. The addition of alarm limits provides a severity reference.

BROADBAND RMS ANALYSIS

The information acquired using broadband RMS data is limited to the total energy that is present in the frequency window for a particular measurement point on a machine-train. Analysis predicated on this type of data is limited to comparing the total energy to a reference data set. In most cases, the broadband RMS data are compared to the initial baseline or to industrial standards. At best, this technique provides a gross value. It will, however, meet the first definition of predictive maintenance. By trending or comparing the current reading to either the baseline data set or industrial standards, you can determine if a problem is present, approximately how severe the problem is, and roughly how long the machine will probably run before failure. You cannot determine either the specific machine component or the failure mode (e.g., imbalance, misalignment) that is present.

NARROWBAND RMS ANALYSIS

Like broadband analysis, narrowband is strictly RMS energy and does not provide any indication of the specific frequency components that make up the total energy. Unlike broadband, narrowband provides the means of monitoring some of the machine-train components. Since narrowband windows are set up around the frequency components generated by specific machine-train components (e.g., gearsets, bladepass), the RMS energy in each window can be directly attributed to a specific machine component. Using the same comparative techniques as broadband analysis, narrowband allows the user to track and evaluate specific machine-train components.

Even though narrowband analysis improves the diagnostic capabilities of your predictive maintenance program, you still do not have the ability to isolate and identify specific failure modes within the machine-train. In addition to the overall RMS, narrowband trends can indicate the relative energy in select machine-train components (e.g., gearbox, bearings). In effect, this type of analysis is a series of mini-overall-RMS readings. The advantage is that specific machine-train components can be monitored. In this manner, you can identify the specific machine components that are degrading, yet you still do not have the ability to determine the failure mode or specific problem.

All microprocessor based predictive maintenance systems should include the capability to automatically trend, alarm, and report all machine-trains that exhibit indications of a change in the vibration signature (FFT) from the established baseline. Typically, after each data set is acquired, the predictive maintenance system will generate a report that identifies every machine-train that has (1) exceeded an alarm limit, (2) has reached an alarm limit, and/or (3) is approaching an alarm limit.

SIGNATURE (FFT) ANALYSIS

Incipient failures in rotating and reciprocating machinery have characteristic frequencies associated with the mode of failure. This phenomenon is due to the fact that the forcing function caused by a developing defect has a unique characteristic signature indicative of that specific defect.

The diagnostic and analysis techniques associated with a predictive maintenance program can be divided into four types: (1) comparative analysis, (2) trending, (3) failure-mode analysis, and (4) root-cause analysis.

If a comprehensive microprocessor-based predictive maintenance system is used, the first two can be automated so little or no manual analysis is required. Such a system, utilizing a properly configured data base, can automatically trend vibration data on each machine-train, compare it to established baselines, and generate trend, time-to-failure, and alert/alarm status reports. Using just the information and capabilities of the automated system, unscheduled failures can be greatly reduced. However, it does not eliminate the root-cause of the premature failure of a machine-train component. Further analysis by the program's staff and/or a consultant can (in most cases) identify the reason for failure and provide recommended corrective actions to prevent a recurrence.

Again, the specific system utilized will determine how much manual effort is required for this analysis. Regardless of the system, the following steps will greatly improve the success of your analysis. It is assumed that additional analysis will only be instituted on suspect machine-trains that, based on the automatic trending, exhibit excessive vibration. Obviously, machine-trains that are operating within acceptable boundaries do not require further investigation. Care should be taken to ensure that the automated function of the predictive maintenance system reports abnormal growth trends as well as machine-trains that are actually in alarm.

COMPARATIVE ANALYSIS

Developing problems within a machine-train can be identified by comparing the signature (FFT) to the baseline signature, previous signatures, or industrial standards. Again, this method does not identify the specific machine component or the failure mode. It can, however, verify that a problem exists.

Figure 12-3 is an example comparing a current signature with previous data, using a strip format. Visual comparison of the signatures enables the user to determine if a problem is developing.

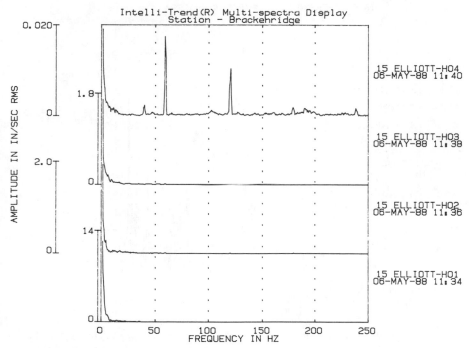

Figure 12-3. Exception reports lists all machines that have exceeded one or more of the alert alarm limits.

Comparative analysis can be used to isolate, within the machine-train, the source of developing problems. By using the common-shaft technique of analysis, the location of the strongest vibration can be determined. By visually comparing the signatures of all measurement points on a common shaft, the source of abnormal vibration can be isolated. Again, this does not identify the problem, but it reduces the number of machine components that must be inspected to correct the problem.

Many of the microprocessor-based predictive maintenance systems allow the user to directly compare the relative strengths of each frequency component. Usually two formats are available: ratio and difference. Figure 12-4 illustrates a direct comparison of the ratio of two spectra; Figure 12-5, is an example of the difference between two spectra.

This type of comparative analysis does not require knowledge of

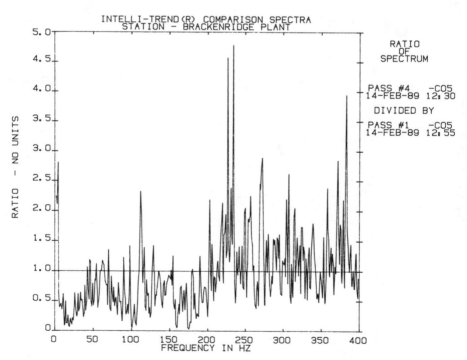

Figure 12-4. Comparative analysis using overlay format helps to visualize actual change in each frequency component.

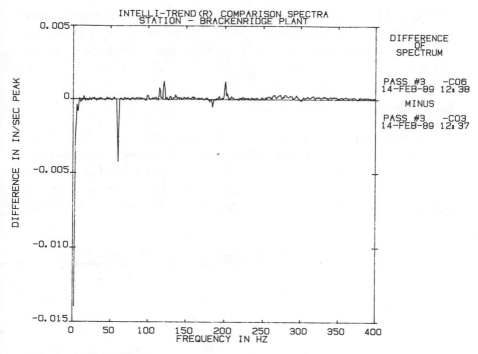

Figure 12-5. Common shaft comparative analysis plots vibration signatures taken from all measurement points on the "common" machine-train shaft. This method helps to isolate the source of a machine problem.

the machine-train or of vibration analysis techniques. As a result, the user cannot determine the specific machine component or the failure mode of problems identified.

TRENDING

Trending of the signatures (FFTs) is another form of comparative analysis. Like RMS trending, this method visually compares the relative change of a machine-train's vibration over time. Unlike RMS trending, trending signatures allow the user to identify the specific frequency components that are creating the change in the vibration amplitude. Trending alone does not identify the specific problem or failure mode of a machine-train, but provides a simple means of determining if a problem exists.

Figure 12-6 is an example of the type of signature trend plots that most microprocessor-based systems can generate. The key to this type of analysis is the visual identification of changes in the signature.

FAILURE-MODE ANALYSIS

All of the analysis techniques discussed to this point have been methods to determine if a problem exists within the machine-train. Failure-mode analysis is intended to identify the specific problem and/or failure mode of the machine-train. Two types of data are required to identify the cause of machine-train problems: (1) vibra-

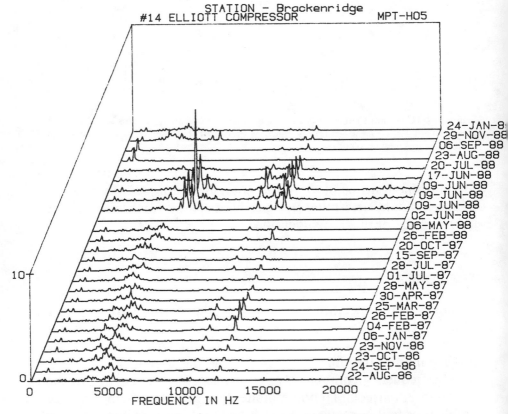

Figure 12-6. Ratio of spectra. Dividing a vibration signature by a reference or baseline provides a measurement of change of each frequency component.

tion signatures (FFTs) from the machine-train and (2) design/operating information on the machine-train.

The premise of failure-mode analysis is that there are established failure modes for machine-trains. Failure-mode analysis also assumes that the vibration pattern for each of these failure modes is identifiable. In general terms, this is true. There are a number of failure-mode charts available to assist the analyst, and they provide an approximation of the problem(s) in a machine-train. An example, Figure 12-7, is the failure mode that states: "Imbalance will generate a dominant fundamental (1X) frequency component with no harmonics (e.g., 2X, 3X)." In many cases, this is true. However, many other machine-train problems will also excite the fundamental frequency component.

Failure-mode analysis can reduce the number of probable causes of a machine-train problem to a workable number. Using this technique, you can identify the most probable causes of a machine-train problem, but you must verify the actual cause by visual inspection, additional testing, or other techniques. Do not assume that because the dominant component is at the fundamental running speed that imbalance is the problem.

Typical Failure Modes

Many of the common defects in machinery components can be identified by understanding their relationship to the true running speed of a shaft within the machine-train. Figure 12-7 identifies some of the more common failure modes.

Charts like Figure 12-7 are general guidelines to the most common failure modes. They cannot provide positive identification of machine-train problems. Such verification requires an understanding of the failure mode and how it looks in the vibration signature. The following sections present general definitions of the most common machine-train failure modes.

Imbalance

Imbalance is probably the most common failure mode in process machinery. The assumption that actual mechanical imbalance must exist to create an imbalanced condition within the machine is incorrect. Aerodynamic or hydraulic instability also can create massive

NATURE OF FAULT	FREQUENCY OF DOMINANT VIBRATION (Hz = rpm 60)	DIRECTION	REMARKS
ROTATING MEMBERS OUT OF BALANCE	1 x rpm	RADIAL	A COMMON CAUSE OF EXCESS VIBRATION IN MACHINERY
MISALIGNMENT & BENT SHAFT	USUALLY 1 x rpm OFTEN 2 x rpm SOMETIMES 3&4 x rpm	RADIAL & AXIAL	A COMMON FAULT
DAMAGED ROLLING ELEMENT BEARINGS (BALL, ROLLER, ETC.)	IMPACT RATES FOR THE INDIVIDUAL BEARING COMPONENTS* ALSO VIBRATIONS AT VERY HIGH FREQUENCIES (20 to 60 kHz)	RADIAL & AXIAL	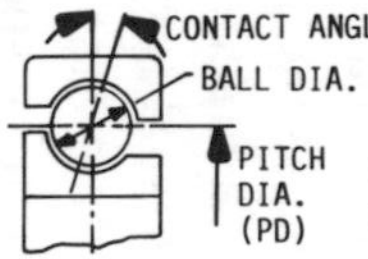 IMPACT RATES f (Hz) FOR INNER RACE DEFECT $f(Hz) \frac{n}{2} f_r (1 + \frac{BD}{PD} \cos \beta)$ FOR OUTER RACE DEFECT $f(Hz) \frac{n}{2} f_r (1 - \frac{BD}{PD} \cos \beta)$ FOR BALL DEFECT f(Hz) $\frac{PD}{BD} f_r (1 - \frac{BD}{PD})^2 \cos \beta^2$ CONTACT ANGLE β BALL DIA. (BD) PITCH DIA. (PD) N = NUMBER OF BALLS OR ROLLERS f_r = RELATIVE REV./S BETWEEN INNER & OUTER RACES
JOURNAL BEARINGS LOOSE IN HOUSINGS	SUB-HARMONICS OF SHAFT rpm, EXACTLY 1/2 or 1/3 x rpm	PRIMARILY RADIAL	LOOSENESS MAY ONLY DEVELOP AT OPERATING SPEED AND TEMPERATURE (eg. TURBOMACHINES)
OIL FILM WHIRL OR WHIP IN JOURNAL BEARINGS	SLIGHTLY LESS THAN HALF SHAFT SPEED (42% to 48%)	PRIMARILY RADIAL	APPLICABLE TO HIGH-SPEED (eg. TURBO) MACHINES
HYSTERESIS WHIRL	SHAFT CRITICAL SPEED	PRIMARILY RADIAL	VIBRATIONS EXCITED WHEN PASSING THROUGH CRITICAL SHAFT SPEED ARE MAINTAINED AT HIGHER SHAFT SPEEDS. CAN SOMETIMES BE CURED BY CHECKING TIGHTNESS OF ROTOR COMPONENTS
DAMAGED OR WORN GEARS	TOOTH MESHING FREQUENCIES (SHAFT rpm x NUMBER OF TEETH) AND HARMONICS	RADIAL & AXIAL	SIDEBANDS AROUND TOOTH MESHING FREQUENCIES INDICATE MODULATION (eg. ECCENTRICITY) AT FREQUENCY CORRESPONDING TO SIDEBAND SPACINGS. NORMALLY ONLY DETECTABLE WITH VERY NARROW-BAND ANALYSIS.
MECHANICAL LOOSENESS	2 x rpm		
FAULTY BELT DRIVE	1,2,3&4 x rpm OF BELT	RADIAL	
UNBALANCED RECIPROCATING FORCES AND COUPLES	1 x rpm AND/OR MULTIPLES FOR HIGHER ORDER UNBALANCE	PRIMARILY RADIAL	
INCREASED TURBULENCE	BLADE & VANE PASSING FREQUENCIES AND HARMONICS	RADIAL & AXIAL	INCREASING LEVELS INDICATE INCREASING TURBULENCE
ELECTRICALLY INDUCED VIBRATIONS	1 x rpm OR 1 or 2 TIMES SYCHRONOUS FREQUENCY	RADIAL & AXIAL	SHOULD DISAPPEAR WHEN TURNING OFF THE POWER

VIBRATION TROUBLE SHOOTING CHART

Figure 12-7. Difference of spectra. Subtracting a reference or baseline signature from a current signature provides a clear indication of change.

imbalance in the machine. In fact, all failure modes will create some form of imbalance in the machine. When all failures are considered, the number of machine problems that are the result of actual imbalance of a rotating element within a machine is relatively small.

Imbalance takes many forms in the signature (FFT). In almost every case, the fundamental (1X) frequency component is excited and is the dominant amplitude. However, this condition can excite multiple harmonics (e.g., 2X, 3X, 4X) depending on the number of planes and their phase relationship.

Misalignment

This condition is virtually always present in machine-trains. We generally assume that misalignment exists between shafts that are connected by a coupling, v-belts, or other intermediate drives. However, misalignment can exist between the bearings of a solid shaft or at other points within the machine.

The presentation of misalignment in the vibration signature depends on the type of misalignment. Figure 12-8 describes three types of misalignment. Each will excite the fundamental (1X) frequency component since they create an imbalanced condition in the machine. The first and second examples will also excite the second (2X) harmonic frequency. The offset created by the misalignment is identical to the second mode of a shaft, and the two high spots created by the shaft turning one revolution creates the second (2X) component. The third example can take several forms. It will excite the fundamental (1X) and secondary (2X) components. It can also excite the third (3X) frequency, depending on the actual phase relationship of the angular misalignment. It will also create a strong axial vibration.

Bent Shaft

A bent shaft creates an imbalanced and/or misaligned condition within the machine-train. Normally, this will excite the fundamental (1X) and secondary (2X) running-speed components in the signature. It is difficult to determine the difference between a bent shaft, misalignment, and imbalance without visual inspection.

Mechanical Looseness

Looseness can create a variety of patterns in the signature. In some cases, a frequency component at ½ the shaft running speed will be

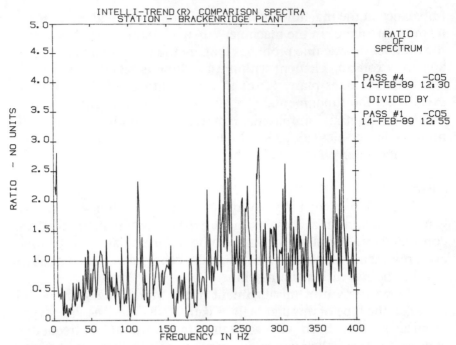

Figure 12-8. Typical signature trend plot or "waterfall" plot can provide a clear picture of change over time. It is an excellent diagnostic tool for variable speed and load applications.

present. At other times, the fundamental (1X) frequency will be excited. In almost all cases, there will be multiple harmonics with almost identical amplitudes present.

Resonance

Every machine-train has a natural frequency. If this frequency is excited by some component of the normal operation of the system, the machine structure will amplify the energy, perhaps causing severe damage. An example of resonance is a tuning fork. If you activate a tuning fork by striking it sharply, it will vibrate rapidly. As long as the tuning fork is held suspended, the vibration will decay with time. However, if you place the tuning fork on a desk top, it will excite the natural frequency of the desk and dramatically amplify the vibration energy. The same thing occurs when one or more of the running speeds in a machine excite the natural structural frequency of the machine or support structure. Resonance is a very destructive vibration. In most cases, it will cause major damage to the machine or support structure.

Natural frequencies of machine-trains and support structures are normally relatively low. In almost all cases, resonance will occur at less than the fundamental (1X) running speed in the signature and will have very high amplitude. One method of determining if an observed frequency component in the signature is a resonance (natural) frequency is to shut off the machine and record the frequency component created when the machine casing is struck with a brass hammer.

Critical Speeds

All machine-trains have one or more critical speeds. If the machine is allowed to operate at any one of these critical speeds, severe vibration and damage will occur.

Each of the criticals has a well-defined vibration pattern. The first critical will excite the fundamental (1X) frequency component; the second critical the secondary (2X) component and the third critical will excite the third (3X) frequency component.

The best way to confirm a critical-speed problem is to change the operating speed of the machine-train. If the problem is critical speed, the amplitude of the vibration component (e.g., 1X, 2X, 3X) will immediately drop when the speed is changed. If the amplitude remains relatively constant when the speed is changed, the problem is not critical speed.

Process Instability

Normally associated with bladed or vaned machinery such as fans or pumps, process instability creates an unstable or imbalanced condition within the machine. In most cases, this will excite the fundamental (1X) and blade-pass/vane-pass frequency components. Unlike true imbalance, the blade-pass/vane-pass frequency components will be broader and have more energy in the form of sidebands. In most cases, this failure mode will also excite the third (3X) frequency component and result in strong axial vibration.

Depending on the severity of the instability and the design of the machine, process instability can also create a variety of shaft mode shapes. This, in turn, will excite the 1X, 2X, and 3X radial vibration components.

Gear Damage

All gear sets create a frequency component (gear-mesh) equal to the number of teeth times the running speed of the shaft. In addition, all gear sets create a series of sidebands on both sides of the primary gear-mesh frequency. In a normal gear set, each of these sidebands is spaced at the running speed (1X) of the shaft, and the profile is symmetrical. See figure 7-9 (page 85) which represents a normal gearset.

If the gear set develops problems, the amplitude of the gear-mesh frequency will increase, and the symmetry of the sidebands will change. The pattern shown in Figure 7-10, on page 86, is typical of a defective gear set. Note the asymmetrical relationship of the sidebands.

Rolling-Element Bearing Problems

There are four normal defect frequencies associated with rolling-element bearings: (1) Fundamental Train Defect Frequency (FTF), (2) Ball-Pass Inner-Race Defect Frequency, (3) Ball-pass Outer-Race Defect Frequency, and (4) Ball-Spin Defect Frequency. Each of the easily calculated frequencies will identify specific defects within a rolling element bearing.

Sleeve (Babbitt) Bearing Problems

Generally instability of the lubricating film is the failure mode for sleeve, or babbitt, bearings. The instability is caused by eccentric rotation of the machine shaft as a result of imbalance, misalignment, or other machine/process problems. When the instability occurs, frequency components at even fractions (e.g., $\frac{1}{4}$, $\frac{1}{3}$, $\frac{3}{8}$) of the fundamental (1X) shaft speed will be excited.

This type of instability is oil whirl or oil whip. As the severity of the instability increases, the frequency components will become more dominant in a band between 0.4 and 0.48 of the fundamental (1X) shaft speed.

ROOT-CAUSE FAILURE ANALYSIS

Diagnostics using the common failure modes is relatively simple; however, the results are not absolute. At best, you can narrow the number of probable causes of observed machine problems. In all cases, further investigation is required to isolate the true root-cause of the problem.

Root-cause failure analysis is the most accurate method of isolating the specific machine component that is degrading and the reason for the degradation.

Traditional predictive maintenance programs typically do not extend beyond the diagnostic phase. Most programs are considered successful if they identify incipient problems in time to prevent severe damage and/or forced downtime. Even with these "successful" programs, two questions remain unanswered: "Why did the problem occur? Will it recur?"

To achieve maximum benefit from a predictive maintenance program, it is not enough to predict failure in time to repair it without affecting production. Machine-train failures do not occur without a reason. In order to solve the problem, the reason for premature failure must be found and corrected. Root-cause failure analysis is the method used to analyze machine-train operation and how they affect the vibration spectrum. Root-cause analysis may have to address the entire process system or systems to determine the reason behind a machine-train failure. If the chronic problem is system related, rather than machine-train related, the knowledge of process dynamics needed for root-cause analysis may be beyond the capabilities of the predictive maintenance staff and require the assistance of the plant engineering staff and/or outside consultants. In extreme cases, the experience of the plant engineering staff will not be sufficient to cost-effectively resolve a chronic problem.

Typical plant engineers, or engineers in general, are generalists within a specific discipline (e.g., mechanical engineering, electrical engineering). They are not, and should not be, specialists in machinery and/or process design and application. Unfortunately, many of the chronic machinery problems that occur every day do require special knowledge of the specific machinery and system to find the solution. In most cases, the first line of support for this expertise is the original equipment manufacturer. The original manufacturer should certainly have the required machine design knowledge; but unfortunately this is not always the case. In most cases, an unbiased consultant can and will provide a cost-effective solution to a chronic problem.

True root-cause analysis is not necessary for many machine-train problems. The failure-mode analysis described in Chapter 7 provides all of the information required to resolve routine, nonrecurring problems. However, if one or more machine-trains exhibit chronic problems, root-cause analysis is essential. A classic example of this is a recurring centrifugal air-compressor problem that a plant recently experienced. In this instance, three 4-stage centrifugal compressors, rated at 2,150 cfm at 100 psig on a common header, exhibited chronic failure problems. In one period of approximately 12 months, all three compressors required major rebuilds and several minor repairs. The predictive maintenance program identified each failure, and the failure analysis isolated and identified the specific failure point; but the problem continued to recur.

A subsequent root-cause analysis indicated that the actual problem was not mechanical (e.g., compressor/motor-related) but was a system (plant air) problem. A redesign of the plant air system eliminated the problem, and normal machine life was restored.

In many cases, recurring machine-train problems can be traced to a system- or process-related problem — not a mechanical (machine-train) problem. Chronic mechanical problems do exist and can usually be traced to improper application, installation, or poor maintenance.

It is difficult to explain the methods utilized in root-cause analysis. The premise of root-cause analysis, like failure-mode analysis, is that there is an identifiable reason or reasons for each abnormal occurrence in machine-train and/or system operation. However, each application is unique, and the methods used vary with the specific process system and/or application. Root-cause analysis requires a complete knowledge of machine and process system design, operation, and typical failure-modes. Many of the chronic problems may be unique to a particular plant, but many have occurred in other applications, plants, or industries. Unfortunately, there is no viable method of communicating this generic failure-mode information between competitive companies, plants, or industries.

Typically, major problems will require the assistance of a consultant with an extensive knowledge of the specific process system and application, as well as a knowledge of typical failure-modes in a variety of industries. Identifying the compressed-air system's problem as outlined above is relatively straightforward — if the analyst's experience base includes previous exposure to generic plant air problems. Without this experience base, the analyst would assume that the problem was machine-train related and continue to investigate only the potential mechanical (compressor) causes.

Root-cause analysis can be expensive, but the net result is always substantial overall savings. In most cases, the cost of a root-cause analysis is less than one major repair and/or rebuild of the machine-train involved.

Some machinery and/or process systems are more susceptible to chronic process- or application-induced problems than are others. Compressors, pumps, and fans are examples of machinery that typically exhibit chronic problems as a direct result of process or application problems. For some reason, these machines are misapplied and/or are extremely sensitive to process variables. Chronic bearing failures are collectively, probably the single biggest example of a machine/system-induced problem. As we discussed earlier, bearings are (by design) the weakest link in most machinery. High failure rates are definite indication of a system/machine problem. Chronic bearing failure on a particular machine or machines should not be ignored. If the bearings fail, other mechanical damage may also be present. Chronic bearing failure is a definite indicator that machine-train life is being adversely affected.

In most instances, these problems can be resolved at a minimal expense, but they usually are ignored and machine failures continue to recur. The net result is that machine-life is drastically shortened, maintenance costs are increased, and available production time is reduced. Ignoring chronic problems, no matter how small they seem, costs a typical plant a substantial amount of actual and/or potential revenue every year.

The methods used in root-cause failure analysis are basically the same as those used for failure-mode analysis. The big difference is that the vibration signature is only one part of the data required to isolate the problem. In addition to vibration data, complete process data (e.g., flow, pressure, temperature), detailed process-system design data, and machine design data are required. In some instances, additional testing utilizing techniques other than vibration analysis will be required. However, the concept of root-cause analysis is to understand the inter-relationship of all of the components within the system or systems so that the real cause of a problem can be identified.

Root-cause analysis techniques, unfortunately, cannot be learned quickly. In addition to formal technical education and general experience, a good experience base in both normal and abnormal systems operation is required to cost-effectively identify and correct most of the chronic problems. As a result, it is more cost-effective to utilize a consultant to resolve chronic problems, rather than train in-house staff.

However, simple root-cause analysis can and should be included in a comprehensive predictive maintenance program. Seminars are available from a number of consulting and/or predictive maintenance systems manufacturers to provide the basic knowledge required for simple root-cause analysis. Simple root-cause analysis is certainly within the capabilities of most predictive maintenance personnel, and with some technical support from outside consultants, most chronic problems can be resolved.

GLOSSARY

Absolute Fault Limits: The maximum recommended level of overall vibration that can be accepted in machinery.

Acceleration: The time rate of change of velocity. Typical units are ft/sec/sec, and g's ($1g = 32.17$ ft/sec/sec $= 9.81$ m/sec/sec). Acceleration measurements are usually made with accelerometers.

Accelerometer: Transducer whose output is directly proportional to acceleration. Most use piezoelectric crystals to produce output.

Aliasing: A phenomenon which can occur whenever a signal is not sampled at greater than twice the maximum frequency component. Causes high-frequency signals to appear at low frequencies. Aliasing is avoided by filtering out signals greater than ½ the sample rate.

Alignment: A condition whereby the axes of machine components are either coincident, parallel, or perpendicular, according to design requirements.

Amount of Unbalance: The weight required in one or more planes to effect coincidence of an object's mass centerline and rotational centerline. Usually given as the product of the weight and the radius at which it must be located.

Amplification Factor (Synchronous): A measure of the susceptibility of a rotor to vibration amplitude when rotational speed is equal to the rotor's natural frequency (implies a flexible rotor). For imbalance-type excitation, synchronous amplification factor is calculated by dividing the amplitude value at the resonant peak by the amplitude value at a speed well above resonance.

Amplitude: The magnitude of dynamic motion or vibration. Amplitude is expressed in terms of "peak-to-peak," "zero-to-peak," or rms. For pure sine waves only, these are related as follows: rms $= 0.707$ times "zero-to-peak"; peak-to-peak $= 2$ times "zero-to-peak."

Analysis Parameters: Vibration-analysis term that identifies specific narrowband frequencies that are associated with machine-train components.

Analysis Techniques: Methods used to analyze machines for system condition. Techniques include: vibration, thermography, tribology, and other destructive and nondestructive methods.

Anti-Aliasing Filter: A low-pass filter designed to filter out frequencies higher than ½ the sample rate in order to prevent aliasing.

Asymmetrical Support: Rotor support system that does not provide uniform restraint in all radial directions. This is typical for most heavy industrial machinery where stiffness in one plane may be substantially different than stiffness in the perpendicular plane. Occurs in bearings by design, or from preloads such as gravity or misalignment.

Asynchronous: Vibration components that are not related to rotating speed.

Attitude Angle: The angle between the direction of steady-state preload through the bearing centerline, and a line drawn between the shaft centerline and the bearing centerline. Applies to fluid-film bearings.

Availability: Actual time that plant machinery and systems are available for production. One measurement of maintenance effectiveness.

Averaging: Digitally averaging several vibration measurements to improve accuracy or to reduce the level of asynchronous components. Refer to the definitions of Root-Mean-Square.

Axial: In the same direction as the shaft centerline.

Axial Position: The average position, or change in position, of a rotor in the axial direction with respect to some fixed reference position. Ideally, the reference is a known position within the thrust-bearing axial clearance or float zone, and the measurement is made with a displacement transducer observing the thrust collar.

Balance: The condition which exists in a part such that no vibratory force is transferred to the supporting bearings of that part due to centrifugal force. Occurs when the mass centerline and rotational centerline of a rotor are coincident.

Balance Resonance Speed: A rotative speed that corresponds to a natural resonance frequency.

Ball Pass Frequency–Inner Race: *See* BPFI.

Ball Pass Frequency–Outer Race: *See* BPFO.

Ball Spin Frequency: *See* BSF.

Band-Pass Filter: A filter with a single transmission band extending from lower to upper cutoff frequencies. The width of the band is determined by the separation of frequencies at which amplitude is attenuated by 3 db (0.707).

Bandwidth: The spacing between frequencies at which a band-pass filter attenuates the signal by 3 db. Window factors are: 1 for uniform, 1.5 for Hanning, and 3.63 for flat top.

Baseline Spectrum: A vibration spectrum taken when a machine is in good operating condition; used as a reference for monitoring and analysis.

Beat Frequency: The periodic change in vibration resulting from the superimposing of two different frequencies. The rate of increase and decrease in signal level is dependent on the difference in frequencies.

BHP: Brake horsepower. The amount of power required by machinery. The amount of energy required to operate machinery.

Blade Pass Frequency: A potential vibration frequency on any bladed machine (turbine, axial compressor, fan, etc.). It is represented by the number of blades times shaft rotating speed (running speed).

Bode: Rectangular coordinate plot of 1X component amplitude and phase (relative to a keyphasor) vs. running speed.

Bow: A shaft condition such that the geometric centerline of the shaft is not straight.

BPFI: Ball pass frequency–inner race. The relative speed between the balls or rollers in a rolling-element bearing and the inner race.

BPFO: Ball pass frequency–outer race. The relative speed between the balls or rollers in a rolling-element bearing and the outer race.

Brake Horsepower: *See* BHP.

Brinneling (False):   Impressions made by rolling element bearings on the bearing race. Typically caused by external vibration when the shaft is stationary.

Broadband Trending: Vibration analysis technique that plots the change in the overall or broadband vibration of a machine-train.

BSF: Ball spin frequency. The frequency created by the balls or rollers in a rolling-element bearing. The actual turning speed of each individual ball or roller in a bearing.

Campbell Diagram: A mathematically constructed diagram used to check for coincidence of vibration sources (e.g., 1X imbalance, 2X misalignment) with a rotor's natural resonances. The form of the diagram is a rectangular plot of resonant frequency (y-axis) vs. excitation frequency (x-axis).

Cavitation: A condition which can occur in liquid-handling machinery (e.g., centrifugal pumps) when system pressure decrease in the suction line and pump inlet lowers fluid pressure and vaporization occurs. The result is a mixed flow which may produce vibration.

Critical Machinery: Machines which are critical to a major part of the plant process.

Critical Speed: The speed of a rotating part which corresponds to its resonance frequency.

Cross-Axis Sensitivity: A measure of off-axis response of velocity and acceleration transducers.

Cycle: One complete sequence of values of a periodic quantity.

Damping: The quality of a mechanical system that restrains the amplitude of motion with each successive cycle. Damping of shaft motion is provided by oil in bearings, seals, etc. The damping process converts mechanical energy to other forms, usually heat.

Decibels (db): A logarithmic representation of amplitude ratio, defined as 20 times the base-ten logarithm of the ratio of the measured amplitude to a reference.

Degrees of Freedom (df): A phrase used in mechanical vibration to describe the complexity of the system. The number of degrees of freedom is the number of independent variables describing the state of a vibrating system.

Discrete Fourier Transform: A procedure for calculating discrete frequency components from sampled time data. Since the frequency domain result is complex, the number of points is equal to half the number of samples.

Displacement: The *actual* change in distance or position of an object relative to a reference.

Displacement Transducer: A transducer whose output is proportional to the distance between it and the measured object.

Dynamic Motion: Vibratory motion of a rotor system caused by mechanisms that are active only when the rotor is turning at a speed above slow roll speed.

Dynamic Range: A measurement of the ability of a data acquisition system (i.e., transducer, signal conditioner, and recording instrument) to capture measurable amplitudes of vibration frequency components. The ability to separate meaningful, low-level frequency components from the noise floor within a machine's signature. Also the ability to simultaneously capture high-amplitude and low-amplitude frequency components within a machine's signature. In normal practice, a minimum of 40 db dynamic range is required for vibration analysis.

Eccentricity, Mechanical: The variation of the outer diameter of a shaft surface when referenced to the true geometric centerline of the shaft.

Eccentricity Ratio: The vector difference between the bearing centerline and the average steady-state centerline.

Eddy Current: Electrical current which is generated (and dissipated) in a conductive material in the presence of an electromagnetic field.

Electrical Runout: An error signal that occurs in eddy current displacement measurements when shaft surface conductivity varies.

Fast Fourier Transform (FFT): A computer (or microprocessor) procedure for calculating discrete frequency components from sampled time data. A special case of the discrete Fourier transform where the number of samples is constrained to a power of 2.

FFT: *See* FAST FOURIER TRANSFORM.

Fluid-Film Bearing: A bearing which supports the shaft on a thin film of oil. The fluid-film layer may be generated by journal rotation (hydrodynamic bearing) or by externally applied pressure (hydrostatic bearing). Typically a sleeve bearing.

Forced Vibration: The oscillation of a system under the action of a forcing function. Typically, forced vibration occurs at the frequency of the exciting force.

Free Vibration: Vibration of a mechanical system following an initial force—typically, at one or more natural frequencies.

Frequency: The repetition rate of a periodic event, usually expressed in cycles per second (Hz), revolutions per minute (rpm), or multiples of rotational speed (orders). Orders are commonly referred to as 1X for rotational speed, 2X for twice rotational speed, etc.

Frequency Response: The amplitude and phase response characteristics of a system.

G: The value of acceleration produced by the force of gravity. One g is equal to 32.17 ft/sec/sec or 9.81 m/sec/sec.

Gear-Mesh Frequency: A potential vibration frequency on any machine that contains gears; equal to the number of teeth multiplied by the rotational frequency (running speed) of the gear shaft.

Hertz (Hz): The unit of frequency represented by cycles per second.

High-Pass Filter: A filter with a transmission band starting at a lower cutoff frequency and extending to (theoretically) infinite frequency.

Hysteresis: Nonuniqueness in the relationship between two variables as a parameter increases or decreases. Also called deadband, or that portion of a system's response where a change in input does not produce a change in output.

Imbalance: Unequal radial weight distribution on a rotor system. A shaft condition such that the mass and shaft geometric centerlines do not coincide.

Impedance, Mechanical: The mechanical properties of a machine system (mass, stiffness, damping, etc.) that determine the response to periodic forcing functions.

Integration: A process producing a result that, when differentiated, yields the original quantity. Integration of acceleration, for example, yields velocity.

Journal: Specific portions of the shaft surface from which rotor-applied loads are transmitted to bearing supports.

Keyphasor: A signal used in rotating machinery measurements, generated by a transducer observing a once-per-revolution event. The keyphasor signal is used in phase measurements for analysis and balancing. Keyphasor is a registered trade name of Bentley-Nevada.

Low Pass Filter: A filter whose transmission band extends from dc to an upper cutoff frequency.

Machine-Train: Term used to describe a total machine, including machine, drive-train, and driver.

Mechanical Runout: An error in measuring the position of the shaft centerline with a displacement probe that is caused by out-of-roundness or surface imperfections.

Micrometer (Micron): One millionth (0.000001) of a meter. One micron equals 0.04 mils.

Mil: One thousandth (0.001) of an inch. One mil equals 25.4 microns.

Modal Analysis: The process of breaking complex vibration into its component modes of vibration, similar to how frequency domain analysis breaks vibration down to component frequencies.

Mode Shape: The resultant deflected shape of a rotor at a specific rotational speed to an applied forcing function. A three-dimensional presentation of rotor lateral deflection along the shaft axis.

Narrowband Trending: Vibration-monitoring technique that plots the vibration energy of specific frequency components generated by machine components such as bearings gears.

Natural Frequency: The natural frequency of free vibration of a system. The frequency at which an undamped system with a single degree of freedom will oscillate upon momentary displacement from its rest position.

Nodal Point: A point of minimum shaft deflection in a specific mode shape. May readily change location along the shaft axis due to variation in residual imbalance or other forcing function, or to changes in restraint such as an increased bearing clearance.

Oil Whirl/Whip: An unstable free vibration whereby a fluid-film bearing has insufficient unit loading. Under this condition, the shaft centerline dynamic motion is usually circular in the direction of rotation. Oil whirl occurs at the oil flow velocity within the bearing, usually 40 to 49% of the shaft speed. Oil whip occurs when the whirl frequency coincides with (and becomes locked to) a shaft resonant frequency. (Oil whirl and whip can occur in any case where a fluid is between two cylindrical surfaces.)

Orbit: The path of the shaft centerline motion during rotation. The orbit is observed with an oscilloscope connected to x- and y-axis displacement transducers. Some microprocessors have the ability to capture and display orbits.

Oscillator-Demodulator: A signal-conditioning device that sends a radio frequency signal to an eddy-current displacement probe, demodulates the probe output, and provides output signal proportional to both the average and dynamic gap distances.

Peak-To-Peak Value: The difference between positive and negative extreme values of a signal or dynamic motion. *See* AMPLITUDE.

Period: The time required for a complete oscillation or for a single cycle of events. The reciprocal of frequency.

Periodic Vibration: Oscillatory motion whose amplitude pattern repeats in time.

Perturbation: Application of a forcing function to a system by means of an external device in order to study the system response characteristics.

Phase: A measurement of the timing relationship between two or more signals, or between a specific vibration component and a once-per-revolution event.

Phase Angle: The angular measurement from the leading edge of vibration component to the following positive peak of the 1X vibration component.

Pickup: *See* TRANSDUCER.

Piezoelectric: Any material which provides a conversion between mechanical and electrical energy. If mechanical stresses are applied on two opposite faces of a piezoelectric crystal, electrical charges appear on some other pair of faces.

Pivotal Balance Resonance: A balance resonance during which shaft motion pivots through the geometric centerline of the rotor, causing a definite zero-axis crossing within the rotor span.

Preload, Bearing: The dimensionless quantity that is typically expressed as a number from zero to one where a preload of zero indicates no bearing load upon the shaft, and one indicates the maximum preload.

Preload, External: Any of several mechanisms that can externally load a bearing, or shaft. This includes "soft" preloads such as process fluids or gravitational forces, as well as "hard" preloads from gear contact forces, misalignment, and rubs.

Probe: *See* TRANSDUCER.

Probe Orientation: The angular relationship of a transducer with respect to bearing housing, shaft, etc. Normally given as a polar coordinate (e.g., 0 = top, vertical; 90 = right, horizontal; 180 = bottom, vertical; 270 = left, horizontal.

Proximity Probe: A noncontacting device which measures the displacement motion or position of an observed surface relative to its mounting location. *See* DISPLACEMENT TRANSDUCER.

Radial: A direction on a machine which is perpendicular to the shaft centerline. Usually refers to the direction of shaft or casing motion or measurement.

Radial Vibration: Shaft dynamic motion or casing vibration which is in a direction perpendicular to the shaft centerline.

Relative Motion: Vibration measured relative to a chosen reference. Proximity probes measure shaft dynamic motion relative to the probe mounting, usually the bearing or bearing housing. Reference can also be to a baseline signature, or reference spectrum.

Repeatability: The ability of a transducer or monitor to reproduce output readings when the same value is applied. The word "accuracy" is often used incorrectly as a synonym for repeatability.

Resistance Temperature Detector: *See* RTD.

Resolution: The smallest change in applied stimulus that will produce a detectable change in the instrument output.

Resonance: The condition of vibration amplitude response caused by a corresponding system sensitivity to a particular forcing frequency.

RMS: *See* ROOT-MEAN-SQUARE.

Rolloff: The condition (and its magnitude) which describes an intentional or desired attenuation at frequencies above or below a certain frequency.

Root-Mean-Square (RMS):   Square root of the arithmetical average of a set of squared instantaneous values. RMS is the closest approximation of total energy of a vibration spectrum.

RTD: An acronym for Resistance Temperature Detector. A sensor that measures temperature and changes in temperature as a function of resistance.

Scale Factor: *See* SENSITIVITY

Seismic Transducer: A transducer that is mounted on the case or housing of a machine-train and measures casing vibration relative to free space. Both accelerometers and velocity transducers are classified as seismic transducers.

Sensitivity: The magnitude of the change in an output signal to a known change in the value of the measurement variable. Also called the scale factor. Typical sensitivity of accelerometers is 100 mV/g. A change of 1 g will increase the millivolt output by 100.

Signal Attenuation: The reduction in magnitude of a signal. Also, the amount of voltage reduction utilized to reduce large electronic signals down to full scale deviation of the instruments.

Signal Conditioner: A device placed between a signal source (e.g., transducer) and a readout instrument to change the signal. Examples: attenuators, preamplifiers, signal convertors, and filters.

Signal Gain: The increase in magnitude of a signal. Also, the amount of voltage amplification utilized to enlarge small electronic signals up to full scale deviation of the instruments.

Signature (FFT):   Term usually applied to the vibration frequency spectrum unique to a particular machine or machine component, system, or subsystem at a specific location and point of time.

Slow Roll Speed: Low rotative speed of machine-train at which dynamic motion effects from imbalance, etc. are negligible.

Spectrum: Presentation of the amplitude on each frequency component within a machine-trains signature (FFT).

Spectrum Plot: An XY plot in which the X axis represents vibration frequency and the Y axis represents vibration amplitude.

Stiffness: The springlike quality of mechanical and hydraulic elements to elastically deform under load. This concept is applicable to shafts, bearings, casings, and support structures. The units normally employed are lbs/inch or newtons/meter.

Strain: The physical deformation, deflection, or change in length resulting from stress. The magnitude of strain is normally expressed in microinches/inch.

Strain Gauge: A transducer that reacts to changes in strain, typically through changes in resistance.

Stress: A force acting on a body per unit area. Usually measured in terms of lbs/inch/inch or newtons/meter/meter.

Subharmonic: Sinusoidal quantity of a frequency that is an integral submultiple of a fundamental frequency.

Subsynchronous: Component of a vibration signal that is less than the fundamental (1X) running speed of the shaft.

Supersynchronous: Component of a vibration signal that has a frequency greater than the fundamental (1X) shaft running speed.

Suppression: The practice of using electronic circuitry to arithmetically subtract the amplitude of an unwanted signal (noise). Not recommended for vibration measurement and/or monitoring because most noise sources are vector, not scalar, quantities.

Sweep Frequency Filter: A type of filter which is automatically swept through a frequency range of interest. An instrument which incorporates this type of filter can be used to generate a vibration frequency spectrum as long as the frequency content of the measured signal remains constant throughout the time required to sweep through the frequency range.

Synchronous: Vibration frequency components that vary in direct proportion to changes in the fundamental (1X) rotative speed.

Tension: A condition in which forces act on a body in line with its axis and cause elongation of the body.

Thermocouple: A temperature-sensing device comprised of two dissimilar metal wires that produce a proportional change in electrical potential at the point where they are joined.

Thermography: Predictive maintenance technique that uses the heat emissions of machinery or plant equipment as a monitoring and diagnostic tool.

Threshold: The smallest change in the measured variable that will result in a measurable change in an output signal. *See* DYNAMIC RANGE.

Thrust: Axial forces and vibration created by the mechanical and/or dynamic operation of machine-trains and/or process systems.

Torque: A moment/force couple applied to a rotor in order to sustain acceleration/load requirements. A twisting load imparted to shafts as the result of induced loads/speeds. Normally requires a simultaneous measurements of radial forces at both ends of a shaft and their phase relationship.

Tortional Vibration: Amplitude modulation of torque measured in degrees peak-to-peak, referenced to the axis of shaft rotation.

Transducer: A device for transmitting the magnitude of one quantity into another quantity. The second quantity often has dimensions different from the first and serves as the source of a useful signal. Vibration transducers convert mechanical motion into a proportional electronic signal.

Transient Vibration: Temporarily sustained vibration of a mechanical system. It may consist of forced or free vibration, or both. Typically, this is associated with changes in machine-operating conditions such as speed or load.

Tribology: The science of rotor-bearing-support system design and operation. Predictive maintenance technique that uses spectrographic, wear particle, ferrography, and other measurements of the lubricating oil as a diagnostic tool.

Trigger: Any event which can be used as a timing reference. A tachometer is typically used to trigger data acquisition to acquire phase vs. amplitude relationship.

Vane-Passing To Frequency: The vibration frequency component(s) created by the vanes/blades of a machine-train. It represents the number of vanes/blades times the fundamental (1X) running speed of the shaft.

Vector: A quantity which has both magnitude and direction. For example, measurements of rotational speed (1X) vibration for balancing will be specified as a magnitude of vibration (mils, ips, etc.) acting in a specific directions (degrees). Most vibration frequency components have a dominant vector (direction). Understanding the vector (direction) of each vibration components and their relationship to the mechanical motions within the machine-train is an aid to diagnosing developing problems.

Velocity: The time rate of change of displacement. This is often expressed as V, or dx/dt. Velocity leads displacement by 90° in time. Typical units for velocity are inches/second or millimeter/second and expressed as peak (zero-peak) amplitudes. Velocity is the relative movement and does not represent the *actual* movement (displacement) of a rotating shaft.

Wear Particle Analysis: A predictive maintenance technique that uses the metallic particles in lubricating oil to determine machine condition.

Index

Index